SYSTEMS AND SYSTEMS THINKING

YAVUZ ERCIL AND
CIGDEM BASKICI

Order this book online at www.trafford.com
or email orders@trafford.com

Most Trafford titles are also available at major online book retailers.

Print information available on the last page.

ISBN: 978-1-4907-9998-8 (sc)
ISBN: 978-1-6987-0000-7 (hc)
ISBN: 978-1-4907-9999-5 (e)

Library of Congress Control Number: 2020904869

Cover Design: Oguz TUNC.

Trafford rev. 06/04/2020

www.trafford.com
North America & international
toll-free: 1 888 232 4444 (USA & Canada)
fax: 812 355 4082

CONTENTS

SYSTEMS SCIENCE

CIGDEM BASKICI

PHILOSOPHIC BACKGROUND
OF SYSTEMS THINKING

YAVUZ ERCIL AND MERVE KADAN

SYSTEMS THINKING BASED
STRATEGY DEVELOPMENT

UFUK TUREN

SYSTEMS THINKING, CYBERNETICS AND VIABLE SYSTEMS MODEL

MEHMET HILMI OZDEMIR

SYSTEM DYNAMICS APPROACH

Yavuz Ercil And Cigdem Baskici

NIKLAS LUHMANN'S THEORY OF SOCIAL SYSTEMS

Elif Ozuz Dagdelen

SYSTEM APPROACH IN THE INTERNATIONAL RELATIONS

KIVILCIM ROMYA BILGIN

ARTIFICIAL INTELLIGENCE AND INTELLIGENT SYSTEMS

ALI SERHAN KOYUNCUGIL AND NERMIN YENIKOSE

TEXT MINING AS A RESEARCH AND DECISION MAKING TOOL IN SYSTEMS

Suat Atan

FOREWORD

I encountered the "systems" concept in 1980s in the context of modern theories of management. Then, I figured out that systems approach is more than a synthesis of different theories. With its emphasis on biological models and its philosophical approach, the first full fledged model that came into my attention was Ludwig von Bertalanffy's "General Systems Theory". Then, as a psychiatrist, Ross Ashby put forward the first design proposals for "Cybernetics and General Theory of Adaptive Systems". The third important contribution to systems understanding came with Jay Forrester's "World Dynamics" model which made a contribution of dynamism and meta modeling to the systems. These three studies were in fact the basic stepstones for me to grasp at least some aspects of the big "Systems Science".

Being provocated by Peter Senge and his "Fifth Discipline", in 1990s, I and my colleagues started to work on "Systems Dynamics" with a theoretical perspective. The practical provocator was. During the same period of time, we were participating in conferences not only on "Systems Dynamics" but also on "Operations Research" in order to be able to contribute to management and organization with quantitative models.

In 2000s, Systems Theory became operationalized as "Systems Thinking", and we figured out the contribution of Emery and Trist to systems thinking in modeling the organizations.

Afterwards, "Systems Engineering" has become the major methodology of fourth generation engineering studies in an interdisciplinary nature with machine learning, software engineering, expert systems, artificial intelligence and so on. At the same time, systems engineering became a master's degree and even undergraduate level program in academic institutions.

These were some milestones of systems theory in my approximately 40 years of academic life. When I received the book and analyzed its content, I was pleased to see that the editors and authors were in complete harmony with my understanding of the important scientific contributions in retrospect. Another nice thing that this book shows us is that the systems concept has not lost its popularity. It was exciting to renew and update myself in systems thinking, as I found out that I missed a number of contemporary studies that challenge systems theory.

I congratulate the authors of each chapter for their very competent and enlighting study and scientific effort. I also want to admit sincerely that I am proud of them as colleagues.

Abdülkadir Varoğlu
Başkent University Ankara

PREFACE

The purpose of this book is to augment a basic reference to define and evaluate the phenomena and events in the world and to perceive them with a system perspective. Nurturing this reference point with the extend of different scientific disciplines would support its explanation and identification ability.

In this book, we tried to substantiate this argument by ligurating different disciplines that make up the system view. In the first part of the book, we tried to define the formation of systems thinking within philosophy and logic. In the following parts, we have defined the forms of system thinking, the basic tools and the usage areas of these tools. We worked on practical applications possible in the departments and an exemplary place for these applications.

The target audience of this book is people who are curious about understanding the world from all areas. We hope that the book will help the reader perceive the world as a system and help view complex systems vitally. The information provided by the authors with all kinds of selection experience and academic background has been carefully selected to bring the reader to this goal.

Good readings…

1

Systems Science

Cigdem Baskici

"Remember, always, that everything you know, and everything everyone knows, is only a model. Get your model out there where it can be viewed. Invite others to challenge your assumptions and add their own."
— Donella H. Meadows

Abstract

The system is a science used to make sense of the world in all areas of life. This section examines the onthology that makes up the system science. In this chapter system science is tried to be defined in three categories. These three categories are discussed to determine the general framework of the system science.

The first is systems thinking. In this first category, the intellectual structure that constitutes the system understanding is examined. Systems thinking is a thinking approach that considers events and structures as parts of a whole system that carries out a certain function, and which attempts to ascribe meaning to systems based on their parts and on the interaction of these parts.

In the second category, systems concepts are defined. Systems concepts are essential for the formation of a system language. Systems concepts are definitions of specific functions or parts of the systems. These definitions help to understand the behaviors of the systems. In this category, some concepts such as open / close systems, linear / nolinear systems, simple / complex system are discussed.

In the third category, system models are discussed. System models are used as a tool to analyze the world. In this category, system dynamics, viable sisytems models and causal relations models are examined.

INTRODUCTION

Systems science is the ordered arrangement of knowledge whose domain of inquiry consists of **thinking, conceptualizing,** and **modeling** the behaviors of the real word in a systematic way. Systems thinking represents the main characteristics of systems (Ackoff, 1974). Systems conceptualization includes the definitions of basic concepts of the systems (Ashby, 1960). System modeling covers the tools which are used to create and test systematic experiments to simulate the real word behaviors (Forrester, 1975).

SYSTEMS THINKING (ST)

Systems thinking is a thinking approach that considers events and structures as parts of a whole system that carries out a certain function, and which attempts ascribe meaning to systems based on their parts and on the interaction of these parts (Senge, 1990). Construction of systems science begins with systems thinking and goes back to the beginnings of 1900's. In 1920s, Ludwig von Bertalanffy developed the concept of **Organismic Biology** using a somewhat Darwinian approach (1950). However, his views did not attract significant attention from the scientific community. In 1937, Bertalanffy presented a report entitled The General Systems Theory (GST) (1937). Although this report also failed to attract much attention at its time, it nevertheless represented one of the most important studies in the field of science. It was not until 1950 that Bertalanffy was able to publish his first article in which he clarified and elaborated his thoughts regarding the General Systems Theory (von Bertalanffy, 1950).

During the 1954 meeting of the American Association for the Advancement of Science, a new society was founded by the biologists L. von Bertalanffy, K. Boulding, A. Repoport, and R. Gerard. This society was initially named the Society for the Advancement of

General Systems Theory, and later renamed as the Society for General Systems Research (von Bertalanffy, 1968).

The General Systems Theory (GST) finds its roots in the concept of Organismic Biology. "In contrast to physical forces such a gravity and electricity, life cannot exist outside of individual entities known as organisms. Each organism is a system, and thus a dynamic order with its own interactions, parts and processes." (von Bertalanffy, 1950).

Both organic and mechanical systems can be the subject of systems thinking. For example, the systems thinking used to describe the relationship between the energy spent and the distance covered by a lizard when it is jumping can similarly be used to describe the relationship between the fuel consumed by a car and its movement.

What matters in this context is not so much the similarity of the **objects** or **cases** being considered, but rather the similarity of **processes** (Churchman, 1968). For example, the point that needs to be considered is the similarity between the exponential growth processes for money and the exponential growth processes of human populations. Therefore, in case money is placed on interest at a yearly rate of 7%, and in case a population similarly grows at a rate of 7% every year, both the sum placed on interest and the population in question will double in 10 years. When determining the behavior of a system, the structure of the system (the order and nature of the processes) is generally as important as its components (Senge et al., 1999).

In addition, both natural and artificial systems can be evaluated within the scope of systems thinking. For example, just as it is possible to explain the functioning of the human body by using systems thinking, the organization and functioning of a business can also be explained by using systems thinking.

According to perspectives based on systems thinking, the structures that constitute life are the result of interactions between parts that possess their own inherent characteristics (Gell-Mann, 1995). Thus, by defining these parts and their interactions, it becomes possible to define life itself. A component, or part, is an entity that has its own characteristics, and which can become associated with other components/parts (Forrester, 1961). When different parts come together or are brought together to perform a certain function, this results in the formation of a whole entity known as the system.

Systems bear and reflect the characteristics of their parts. Depending on their function, they will also have their own inherent characteristics. Different systems can also be assembled together to form different and even more complex systems (Flood, 1989).

From this perspective, many of the entities encountered in our lives are systems. Entities that are not systems are referred to as masses. The main difference between masses and systems is the way in which they perform their functions. A mass is limited to its constituting parts. As such, the interactions between these parts do not create a special function. For example, masses of sand, water, or cement are formed by grains of sand, droplets of water, and grains of cement (Ashby, 1962). However, since these constituents do not have a particular function, the characteristics of masses will not change regardless of many parts the initial masses are divided into. If one divides a sand pile into two parts, and then into four parts, one will obtain masses which, despite their smaller sizes, will have the same characteristics as the initial mass. On the other hand, in case sand, water, and cement were brought together within a building system, the relationship between these parts will create not only a new function, but also a new system (Flood & Jackson, 1996). In such a case, dividing the resulting system into smaller parts will cause the system to lose its ability to fulfill its function. For this reason, a system represents a whole that is more than the sum of its parts.

The relationships between the parts that constitute a system are causal with regards to events, and symbiotic with regards to structure. A causal relationship refers to the cause and effect relationship between the interacting parts (Kim, 1994). A causal relationship applies when, within the context of an event, a cause in one part of the system creates an effect at another part of the system. A **symbiotic relationship** refers to the cases when there is a mutual interdependence in the relationships of the different parts constituting the system (Forrester, 1971). With regards to systems science, the basic principles that determine the characteristic of the relationships between the different parts are as follows (Emery, 1969):

1. The relationships are of a dynamic character: Relationships constantly evolve to the state in which the system's function can be best fulfilled.
2. The relationship between the different parts can involve the flow of materials, information, or energy. The same rules are applicable under different circumstances and conditions.

3. The relationships are sensitive to environmental changes.
4. Together, the system's relationships form a dynamic relationship.
5. The system relationships are based on cause and effect relationships.

Relationships indicate how one part can affect another part within the same system. For example, increasing urbanization leads to a decrease in agriculture. Excitement causes blood pressure to rise. In system relationships, the first variable influences the second variable (Shannon, 1964).

In such a relationship structure, it is possible to describe two different relationship groups. The first relationship group refers to the group of relationships between the parts that constitute the system. For systems, the first group relationships represent input-output relationships. All the values that systems require to fulfill their own functions are considered as inputs. On the other hand, the products obtained following the processes and functions of these systems represent the outputs.

SYSTEMS CONCEPTS (SC)

Systems concepts are definitions of specific functions or parts of the systems. These definitions help to understand the behaviors of the systems.

In case a system does not have an input-output relationship with its environment, this system is referred to as a **closed system**. Systems that have input-output relationships with their environment, on the other hand, are known as **open systems** (Sterman, 2001). Within the scope of these relationships, the structure that exists between the inputs and outputs of the system will indicate whether the first group relationships are linear. To determine whether the relationships within this group are linear, two tests are performed (Sterman, 2002).

1. **Homogeneity Test:** In case a change in one of the inputs of the system causes an equal change in the system's outputs, then the first group relationships of the system are considered to be **linear**.

2. **Additivity Test:** In case the sum of the effects of the two inputs is equal to the sum of the effects caused by these inputs through the system, the system is considered to exhibit a linear relationship. However, in case the sum of the effects of the two inputs is different from the sum of the effects caused by these inputs through the system, the system is considered to exhibit a **nonlinear** relationship.

Systems with fully linear relationships are known as **linear systems**, while systems without linear relationships are as **nonlinear systems**.

The other relationships of a system include the relationships between its own parts. In case the changes in the relationships between a system's parts result in predictable and definable changes in the system's behavior, the system is considered to display simple relationships. Systems with such relationships are known as **simple systems** (Reggia, Armentrout, Chou, & Peng, 1993). In case the changes in the relationships between a system's parts result in unpredictable and indefinable changes in the system's behavior, the system is considered to display complex relationships. Systems with such relationships are known as **complex systems** (Bar-Yam, 2019).

Ross Ashby's Law of Requisite Variety (1956) demonstrated that a complex system can only be controlled by another system that possesses a mechanism of equal complexity. According to Ashby's approach, "only variety absorbs variety." More complex systems are more sensitive to their environments. The minimum variety necessary to perceive a complex environment is also determined to be a system's complexity. Social organizations are more complex than physical organizations/systems.

A system undergoes an adaptation process that depends on the structure and nature of the environment in which it is found. This structure is dynamic. As result, part of the change observed in the system will stem from its efforts to adapt to the new environment, while the remaining changes will stem from the attitude and behavior of other systems. This interactive aspect of the environment is reminiscent of the processes that biological systems undergo. Within this structure, the survival efforts of systems are limited to their possible behaviors. The choices of a system within changing structures will determine its likelihood of surviving. These types of environments are called **fitness landscapes**, and are very well known from the study of biological systems (Holland, 1995). For this reason, the first studies and definitions for fitness landscapes have been defined and developed many years ago (first by Herbert Spencer on 1864).

The fitness landscape is composed of the probabilities for different decisions that the organization might make during a change process. Kaufman previously demonstrated that this landscape could be defined by using the number of these probabilities (N) and the **connectivity** of each choice (K) (Kauffman & Weinberger, 1989). Kaufman's work thus allowed the fitness landscape model, which is also known as the NK model, to acquire a mathematical expression. Each choice probability has a corresponding fitness value. This value (which is the mean of all probabilities), describes the area in which a choice is located within the system, and with respect to other choices (i.e. its distance and relationship to other choices). Kauffman defined the fitness landscape mainly for biological systems (Kauffman & Weinberger, 1989).

Self-organization increases when systems interact with their environments at the **edge of chaos** (Gleick, 1987). The flexibility provided by self-organization renders it easier for successful organizations to constantly adapt to the changes in their environment. Three important mechanisms that allow systems to self-organize are identification, information, and relationships (Peters, 1987). A system formed by self-organization acquires certain characteristics and behavior, as well as certain spontaneous properties.

In case a system is a closed system, its function (which represents its reason for existence) will not be dependent or affected by its interactions with the environment (Geiger, 1968). In other words, the system's functions will not be dependent on inputs and outputs. In such a system, the outputs will gradually decrease if the inputs remain the same. This, in turn, will cause the system's energy to constantly decrease. For this reason, closed systems eventually lose the reasons that allowed their constituent parts to remain together, which leads them to fall apart. This tendency of closed systems to fall apart is defined as entropy.

Entropy reflects the fact that the functions of systems are not associated with their environmental interactions, and that they consequently experience a constant loss of energy that ultimately leads them to fall apart (Waldrop, 1992). Thus, to strengthen the bonds between the parts of a system, it is necessary to strengthen the relationship between its function and its inputs-outputs. As such, it is necessary to increase the inputs or the outputs relating to the function. The flow of energy that leads to an increase in the system's inputs or outputs is called **negative entropy** (Ayres, 1988).

To be able to perform their functions in a sustainable manner, the systems attempt to increase their input-output relationships. This allows them to continue their existence – or to continue living. In this context, **living systems** are systems that can constantly develop negative entropy (Weiner, 1961).

The efforts demonstrated by systems in order to increase their negative entropy are known as **adaptation**. Systems that can undergo change in order to develop their input and output relationships are called **adaptive systems** (Narendra, 2013).

The relationships and interactions of systems in both groups are bidirectional. The fact that these relationships are bidirectional means that both the relationships between the system and its parts and the relationships between the system and its environment will have an effect on the system. These effects can reshape the behavior of the system with respect to its activities. This type of interaction is called **feeding**.

Feeding that stems from the relationship between the environment and the system is called **feedforward**. Feedforward involves an evaluation process that progresses from system inputs to system outputs (from backward to forward) (Kreitner, 1982). Before beginning its functions, a system first attempts to obtain input from the environment that are suitable for its functions. This allows the system to shape itself according to future conditions when developing its functions.

Feeding that stems from the relationships between the system and its parts is called **feedback**. Feedback involves an evaluation of whether the system acts according to its functions, through assessments that progress from the system's outputs to the system's inputs (from forward to backward) (Kreitner, 1982).

In case a system function deviates from its previously determined goals, it becomes necessary to perform a correction in the inverse ratio and direction of the deviation (Forrester, 1994). In case this correction increases/improves the function of the system, the feedback is described as **positive feedback**. In case this correction decreases/

reduces the function of the system, the feedback is described as **negative feedback**.

Feedback allows systems to reach a state of equilibrium. An important feature of systems that can reach equilibrium is their ability to react actively to change. Most living systems continue to function and to use energy even when they are not required to display any reactions. Mechanical systems can be entirely deactivated for a certain period of time, and then be reactivated. The same type of behavior is also observed in certain simple life forms. Certain plant roots can wait for years until suitable conditions for growth become available, while certain microorganisms can be frozen and then reanimated. However, most organisms need to remain active in order to stay alive (von Bertalanffy, 1968). In other words, they will inevitably die if they become inactive.

As a general rule, a more complex system will also have a higher level of internal activity. This increases the likelihood of the system to trigger change in its environment. Complex systems spend more energy to process information and to ensure their continuity (Prigogine & Stengers, 1984). This same rule is also applicable for social systems. For example, while a club may be easily closed and then reopened, doing so for a company is far more difficult. A society, on the other hand, needs to be constantly active and to ensure its own continuity.

The behavior of systems is largely determined by their feedback reactions. This phenomenon is called the **reaction time**. The functioning of a simple thermostat/oven involves a long reaction time, since air conveys heat from the oven to the thermostat at a relatively slow rate. Social systems also have generally long reaction times (Weiner, 1998). If the reaction time is too slow, external changes will take place at a comparatively faster rate, such that the system will be unable to respond. In addition to this, the reaction time is generally the same as the basic tempo, or the "shot" of the system (the minimum interval between the oscillations of the feedback process). By watching these oscillations and estimating the time interval, it is possible to determine the minimum time interval that a certain system may overcome.

Forming the decisions of a system also requires its identity to be formed/created. The property of a system that allows sufficient free movement for the formation of an identity is referred to as its **viability** (Beer, 1989).

Systems Models (SM)

Systems models are tools which are representations of real world in different perspectives. There are different types of systems models based on their perspectives. Some of them are explained briefly as follows.

Complex Adaptive Systems Model

Holland (1995) argued that integrating systemic complexity and environmental complexity with one another could only be achieved based on the inherent abilities of the system. Foremost among these system abilities is the adaptation skill. For this reason, such systems are referred to as Complex Adaptive Systems (CAS). The general structure of the CAS is framed according to seven basic laws. Four of these laws describe the characteristics of the CAS. These are the laws on aggregation, non-linearity, flows, and diversity. The other three laws describe the mechanisms of the system, which are tagging, internal models/schemas, and building blocks. CASes can be described as the fundamental elements of the concept of complexity. CASes are self-organizing systems. Self-organizing systems tend to organize themselves by using energy, materials, and feedback from their internal and external environments. Similar to traffic, crowdedness and organization, self-organization is also a characteristic of a multi-component system. Self-administering teams within organizations represent examples of self-organization.

Social Systems Model

Luhmann (1982) contributes to systems science in the area of sociology by constructing the Social Systems Model. Luhmann's views were largely based on the structural functional systems theory, whose foundations established by Parsons (1951). Parsons associates the existence of a system to its function. For the continuity of the function, systems establish their own optimal balances. To ensure balance, it is important to closely follow the system's relationships with the environment, and to ensure that the parts of the system remain together as a whole. It is for this reason that institutions are necessary. With this type of structure, all events that occur within society are the result of interactions between the actors. The action that forms the basis of this interaction harbors four basic elements/aspects: adaptation, goal access, integration, and implicit pattern formation (AGIL).

Luhmann systems are divided into two basic groups, which are the autopoietic and the allopoietic groups. Autopoietic group systems are systems that undergo structural changes of their own when establishing a dynamic balance. These include social (interaction, organization, and social) systems as well as psychological systems. Allopoietic systems, on the other hand, refer to machines and organisms (Midgley, 2003).

System Dynamics Model

Towards the end of the 1950s, Professor Jay W. Forrester (1975) developed a new theory based on concepts relating to feedback, work evaluation, organizational and social relationships, and the structural changes associated with these relationships. The theory involved the modeling of mathematical changes in structural movements.

The rationale of the System Dynamics Modeling is based on the view that the movements or changes triggered in real structures will actually consist of small-scale movements, while complex situations mainly stem from relationships and interactions between the elements of the system (Richmond, 1994). The interaction of all effective system elements on any structure will lead to complexity. It is believed that this complexity can be understood through the analysis of feedback structures that are inherent to all relationships. For this reason, feedback structures play a key role for understanding complexity.

In this context, the graphic illustration and description of System Dynamics Model indicates that this model is very similar to the General Systems Theory. In addition to being easy to use, the system dynamics model also allows the modeling of non-linear relationships within an organization. It thus represents a tool that can be easily used and understood by those in decision-making positions. Although models developed using System Dynamics involve numerous elements and relationships, they do not take much time to prepare, and are quite descriptive (Meadows, Dennis, & Meadows, 1974).

The time-dependent change observed in the elements or variables that constitute a system represent the behavior of a system. The behavior of a system is intrinsic, dependent on the baseline, and non-linear (Stacey, 1992). Systems demonstrate their behavior under the influence of their own structure and elements. This behavior is displayed within the limits and effects of the system's structure. Based on this consideration, it is possible to show that a system's structure also determines the behavior of the system. Thus, System Dynamics identifies the relationship between a system's behavior and the structure that is at the source of this behavior. In other words, System Dynamics evaluates how the structure of a physical, biological, or literary system shapes the displayed behavior of the system (Sterman, 2001). For example, by identifying the structure of an ecological system, System Dynamics allows the time-dependent behavior of this system to be effectively analyzed.

Viable Systems Model

Viable Systems Modeling is constructed by Stafford Beer (1985). It represents a cybernetic approach to systems science. According to Beer, freedom does not define a state that is distant or dissociated from the other entities in the environment. For example, what define a university are the education it provides and the research it performs. Without these functions, an institution would not be considered as a university. As such, all of the units, faculties, departments, centers, etc... that provide education, conduct research, and bring life to the university through their products/outputs can be considered living systems. The other structures within the university such as libraries, dormitories, dining halls, and councils and committees, on the other hand, are not living systems. For a system to be considered a living system, it must be able to interact in autonomously with its environment. Autonomy allows a system to make decisions based on its own abilities/capacities. Viable systems have mission statements, as well as accountabilities within the scope of the functions they carry out in accordance with the policies and strategies of the larger entity of which they are part. The continued viability of systems is achieved by two main mechanisms (Beer, 1984): cohesion, and adaptation. The cohesion mechanism aims to ensure that the goals and interests of the independent units within the main system are in agreement with the policies, strategies, goals, and interests of the main system, and that the competitive strength of these units are reinforced.

The adaptation mechanism, on the other hand, aims to ensure that the system is able to adapt in a timely manner to any changes that might occur in the external environment. In addition to adaptation, learning, development, and evolution are also considered among the basic dynamics of the concept of viability. It would not be possible to achieve evolution without first achieving adaptation; to achieve adaptation without first achieving learning; and to achieve learning without first achieving equilibrium.

Causal Relations Model

Causal Relations Models are system models which represent systems behavior with causal relations. Causal relations do not refer a ranking but refer influences on system functions. The relationship among the system is not linear but circular. The circular definition of system behavior is called as **structure** (Senge, 2014). Different real world behaviors can be modelled with the same structure. In this perspective it is accepted that structures define the behaviors.

A systems structure can be constructed two different kinds of loops. First kind of loops is reinforcing loops. This kind of loops generates either a continuous increase or decrease in systems behavior. For example, a number of customers are increased by increasing the investment of products or services. Increase in number of customers leads to increase in revenues which lead to increase in investments. The other type of loops is balancing loops. In balancing loops at least one (or in odd numbers of) relation is negative relation. For example, if the positive perception of a service quality is increased more people are attracted and it causeS a queue. While the number of the people waiting in the queue is increasing, the waiting time is increased. Increase in waiting time leads to decrease in the perceived service quality. Decrease in perception about service quality leads to decrease in the number of the customers attracted. This loop models a systems behavior which is opposite to the initial behavior.

In causal relations models it is accepted that delays in relations and mental models of the actors can create the changes in systems behaviors. Mental models that cause change in relation are represented in clouds. Delays are represented by two parallel lines.

REFERENCES

Ackoff, R. (1974). *Redesigning the future: A systems approach to societal problems.* NY: John Wiley & Sons, Inc.

Ashby, W. R. (1956). *An introduction to cybernetics.* London: Chapman and Hall.

Ashby, W. R. (1960). *Design for a brain: The origin of adaptive behavior.* London: Chapman and Hall.

Ashby, W. R. (1962). Principles of the self-organizing system. In H. von Foerster, & G. Zopf (Eds.), *Principles of self-organization* (pp. 255-278). New York: Pergamon Press.

Ayres, R. U. (1988). *Self-organization in biology and economics* (IIASA Research Report-January). International Institute for Applied Systems Analysis, Laxenburg, Austria.

Bar-Yam, Y. (2019). *Dynamics of complex systems.* CRC Press.

Beer, S. (1984). The viable system model: Its provenance, development, methodology and pathology. *Journal of the Operational Research Society, 35*(1), 7-25.

Beer, S. (1985). *Diagnosing the system for organizations.* Chichester: John Wiley & Sons.

Beer, S. (1989). Viable system model. In R. Espego, & R. Harnden (Eds.), *Viable systems model.* Chicester: Wiley.

Churchman, C. W. (1968). *The systems approach.* NY: Dell Publishing.

Emery, F. E. (1969). *Systems thinking: Selected readings.* NY: Penguin Books.

Flood, R. L. (1989). Six scenarios for the future of systems "problem solving". *Systems Practice,* (2), 75-99.

Flood, R. L., & Jackson, M. C. (1996). *Critical systems thinking: Directed readings.* NY: John Wiley and Sons.

Forrester, J. W. (1961). *Industrial dynamics.* Cambridge: MIT Press.

Forrester, J. W. (1971). Counterintuitive behavior of social systems. *Technology Review, 73*(3), 53-68.

Forrester, J. W. (1975). *Collected papers of Jay W. Forrester.* Norwalk: Productivity Press.

Forrester, J. W. (1994). System dynamics, systems thinking, and soft OR. *System Dynamics Review, 10*(2-3), 245-256.

Geiger, D. (1968). Closed systems of functions and predicates. *Pacific Journal of Mathematics, 27*(1), 95-100.

Gell-Mann, M. (1995). *The quark and the jaguar: Adventures in the simple and the complex.* NY: W.H. Freeman and Co.

Gleick, J. (1987). *Chaos: Making a new science.* NY: Viking.

Holland, H. J. (1995). *Hidden order: How adaptation builds complexity.* Reading: Perseus Books.

Kauffman, S. A., & Weinberger, E. D. (1989). The NK model of rugged fitness landscapes and its application to maturation of the immune response. *Journal of Theoretical Biology, 141*(2), 211-245.

Kim, D. (1994). *Systems thinking tools: A user's reference guide.* Cambridge: Pegasus Communications.

Kreitner, R. (1982). The feedforward and feedback control of job performance through organizational behavior management (OBM). *Journal of Organizational Behavior Management, 3*(3), 3-20.

Luhmann, N. (1982). *The differentiation of society* (S. Holmes, & C. Larmore, Trans.). New York: Columbia University Press.

Meadows, L., Dennis, B. H., & Meadows, D. (1974). *Dynamics of growth in a finite world*. Cambridge: Wright-Allen Press.

Midgley, G. (2003). *Systems thinking, Volume I&II: General systems theory, cybernetics and complexity*. London: Sage Publications.

Narendra, K. S. (2013). *Adaptive and learning systems: Theory and applications*. Springer Science & Business Media.

Parsons, T. (1951). *The social system*. London: Free Press.

Peters, T. J. (1987). *Thriving on chaos*. NY: Knopf.

Prigogine, R. I., & Stengers, I. (1984). *Order out of chaos*. NY: Bantam Books.

Reggia, J. A., Armentrout, S. L., Chou, H. H., & Peng, Y. (1993). Simple systems that exhibit self-directed replication. *Science, 259*(5099), 1282-1287.

Richmond, B. (1994). Systems thinking system dynamics - Lets just get on with it. *System Dynamics Review, 10*(2-3), 135-157.

Senge, M. P. (1990). *The fifth discipline: The art and practice of the learning organization*. NY: Currency Doubleday.

Senge, M. P. (2014). *The fifth discipline fieldbook: Strategies and tools for building a learning organization*. Crown Business.

Senge, M. P., Kleiner, A., Roberts, C., Ross, R., Roth, G., Smith, B., & Guman, E. C. (1999). *The dance of change: The challenges to sustaining momentum in learning organizations*. NY: Oubleday/Currency.

Shannon, C. (1964). *The mathematical theory of communication*. Urbana: University of Illinois Press.

Stacey, D. R. (1992). *Managing the unknowable: Strategic boundaries between order and chaos*. San Francisco: Jossey-Bass.

Sterman, D. J. (2001). System dynamics modeling: Tools for learning in a complex world. *California Management Review, 43*(4), 8-25.

Sterman, D. J. (2002). All models are wrong: Reflections on becoming a systems scientist. *System Dynamics Review, 18*(4), 501-531.

von Bertalanffy, L. (1937). *Das gefüge des lebens*. BG Teubner.

von Bertalanffy, L. (1950). An outline of general system theory. *British Journal for the Philosophy of Science, 1*, 134-165.

von Bertalanffy, L. (1968). *General system theory: Foundations, development, applications*. New York: Braziller.

Waldrop, M. M. (1992). *Complexity: The emerging science at the edge of order and chaos*. NY: Simon & Schuster.

Weiner, N. (1961). *Cybernetics or control and communication in the animal machine*. Cambridge: The MIT Press.

Weiner, N. (1998). *The human use of humans: Cybernetics and society*. NY: Da Capo Press.

About Author(s)

CIGDEM BASKICI is assistant professor in the Departmant of Health Management at Başkent University. She completed her undergraduate education at Ankara University, Faculty of Political Sciences, Department of Economics. She received her Ms and PhD degrees from Ankara University, Institute of Social Sciences, Department of Business Administration. She works in the fields of international business, strategic management and network theory. Since 2017, she is a member of the board of directors of Başkent University Center for Strategy and Technology. She worked as a researcher in European Union and World Bank projects, and took part in projects carried out in cooperation with university-public institutions.

2

Philosophic Background Of Systems Thinking

Yavuz Ercil And Merve Kadan

> *"We have created trouble for ourselves in organizations by confusing control with order."*
> *– Margaret J. Wheatley*

Abstract

In this section, the intellectual foundations of systems thinking are discussed. For any intellectual approach to be called a system approach, it must have its own intellectual patterns.

Systems thinking is the cognitive structure that reflects these patterns. Three main concepts indice the basic features that distinguish system thought from other thought structures holism, dynamism and causality. It is possible to say that the intellectual structures developed on these basic concepts fall within the scope of systems thinking.

Holism is a whole different from the sum of its parts. All elements in a holistic structure are interdependent and sensitive to the external environment. The speed or shape of environmental changes also affect the elements within the structure.

Dynamism is when the system changes state over time and depending on the starting point. In an effort to understand the behavior of dynamic systems, besides the result of the behavior, the quality of the actuator that initiates the behavior also plays an important role.

Since there is a proportion between inputs and outputs, the behavior of the linear system can be estimated and can therefore be explained in causality. On the other hand, nonlinear systems are irregular, unpredictable and unexplainable. In nonlinear systems, there is no linear proportion between input and output, or cause and effect. The logical structure of systems thinking is formed within these paradigms.

INTRODUCTION

This chapter discusses the intellectual bases of systems thinking. In order for any intellectual approach to be considered systems approach, it has to have specific intellectual patterns. Systems thinking is the cognitive structure that reflects these patterns. Common points of different studies (Brandstädter, Harms, & Grossschedl, 2012; Davidz & Nightingale, 2008; Davis & Stroink, 2016; Hung, 2008; Mulej et al., 2004) on the basic features that distinguish system thinking from other thinking structures can be summarized as holism, dynamism, and causality. It is possible to say that the intellectual structures developed on the grounds of these basic concepts are within the scope of systems thinking. Then, it should be put that the logical structure of systems thinking has formed within the paradigms of these basic variables.

HOLISM AND ENTROPY

The concept of holism was coined by Jan Smuts (1926), who defined it as "the evolutionary whole that is different than the sum of its parts" (142-144). In this case, holism should be defined based on the structure that associates parts with one another. The attraction (or distribution) among parts can be measured by entropy (Sirtes & Oberheim, 2006). All elements within a holistic structure are dependent on each other and sensitive to the external environment. Speed or mode of environmental changes affect the elements within the system, as well. The structure, in an environment in which its change accelerates depending on internal and external factors, can be perceived as an indicator reflecting the impacts of changes.

Under the pressure of internal and external conditions, every mobilization may be the start of a differentiation. In this case, following the opportunity to detect and affect the mobilization within the structure is shaping holism. It is possible to monitor the actual status of the structure and reach interpretations through information obtained with the movement-based approach (Nohira & Berkley, 1994: 75-81) which is thus defined.

Such view of the structure acknowledges that, along with the change of structural balances, the structure must too change. It is possible, within this framework, that states of systems with a simple structure can be defined with ongoing studies. Particularly, within the framework of thermodynamics[1], in which intensified efforts are seen toward defining simple systems by drawing upon the movements of the boundary, the effort is to give a definition with five basic concepts. Knowing any three of these concepts gives information about the dimensions of the others. These basic concepts are listed as **pressure, heat, volume, energy** and **entropy**. Energy, whose sum is constant within the universe, can emerge in various ways and transform from one form into another depending on different values in basic concepts. For example, a substance can be given a new form by changing its volume.

While energy changes from one form to another, heat is released. The released heat cannot be recovered. This irreversibility, called the second law of thermodynamics, creates disorder by repeating for every change. Thus, a picture emerges, where systems change toward constant disorder within a unidirectional transformation. As structural complexity of systems increases, the expression of dissipation in question gets harder. With the first study for defining the dissipation in question (Boltzman, 1995), probabilistic expressions started to be used for defining structural change. Consequently, the understanding that simple systems will have behaviors with different probabilities as they get more complex began to get evident.

Maxwell (1871: 48-53) uses the two glass jars experiment for defining entropy. Accordingly, if glass jars connected to each other by only one channel are observed for a certain period of time, the randomness (different behavior of each molecule with a probability

---

1   For detailed information, see Planck, M (1945) and Maxwell (1871)

with very difficult predictability) of the dissipation of air molecules within them will be determined. If a virtual genie on the lid of one jar that opens to the outside opens the lid and let new molecules enter the jar, then complexity will increase inside the jar but decrease in the external environment. And a reverse situation can be interpreted as the decrease of complexity within the jar.

With this definition, systems may be thought to have behaviors with the same basis and different probabilities, from simple toward more complex. Just as Maxwell imagined genies when defining entropy, all inputs entering an organization may change the organization's structure, as well. This change may be random according to the way the energy coming from outside is assessed within the organization. And it will not always be possible to express this randomness.

It is impossible to consider the units deemed important in the organization, and in fact the whole structure the same between two enterprises, one, which previously played a leading role in a market where competition conditions were not developed enough, choosing a new product strategy in time for protecting its market share against new competitors and products, and the other enterprise in the same situation employing new marketing techniques. This is different for every organization, just like the movements of the molecules in the jar. In an organization, too, every new formation shows a distribution statistically, and it is possible to measure the number of its probabilities.

On this ground, the most distinct way to eliminate the dissipation caused by entropy, that is, to express entropy with negative value is provide new inputs from the environment. The most distinct indicator of negative entropy structurally is information. When changes spread from simple toward complex, in line with the general system behavior, it is possible to mention full irreversibility of changes. If a value can be defined as a probability for each change, the change situations in question form a "set of change probabilities". If these change situations are defined with e, a set of probabilities such as

$$E= \{e1, e2, e3, e4.......en\}$$

can be formed. If realization probabilities of each situation taking a place in this set, which are different from each other, are taken into consideration, the set of realization probabilities measured randomly is expressed as follows:

$$P = \{p1, \ p2, \ p3, \ p4.........pn\}$$

Here, a point that requires attention is that total probabilities equal 1 and each is greater than zero individually. In the light of this information, the realization value of total probabilities can be formulated as follows:

$$H = -K^n \sum_{i=1}^{n} pi \ \log(pi)$$

This is the formula of entropy, which is the second law of thermodynamics, and can be used for any environment where there is uncertainty (Ruelle, 1988: 195-196).

Differentiation can be defined by monitoring the change creation process of a unit (mobilizer) that creates the movement within the structure. Thus, different scales can be achieved in accordance with the mobilizer's structure. For example, change in a company as a social system can be monitored by money flow (Hershey, 1991: 101). In addition, it can also be expressed as information flow (Yager, 1992: 352). Traditionally, it can be expressed as the numerical change of physical measurements (Esen, 1985). In this case, it is required to consider the structure's foundation the starting point. After this point, change will begin. At such a point, it can be thought that the structure is symmetrical, that is, has no mobilizer. The structure's deviations from its symmetrical state will be the expression of change. Then, a concept (symmetry number) will emerge, in which the structure includes all equilibria and this can be expressed by a value. Daniel Hershey (1991: 102) interpreted the symmetry number as a function of structural power. Hershey expressed the entropic definition of integration, emphasizing that it is an outcome of the power within the structure, as follows:

$$SN =^n \sum_{i=1}^{n} Li(Pi - [Pi])$$
$$[Pi] = \sum_{i=1}^{n} p_i / n$$

Pi/n = Average power of the organization

Pi = Power of each unit in the organization
Li = Rank number of the organization (For example, 1 manager rank, 2 deputy ranks)
n = Total number of units in the organization

Here, power refers to the result obtained from the division of organizational value by unit rank number. Daniel Hershey (1991) found by using this information that the symmetry number is -30 for structures without entropy. This means that all power is in a single hand. Thus, environments without entropy appear as settings with high centralization. In the case of 30, which is the maximum symmetry number, it was found that entropy too is maximum. This indicates that all power is distributed.

Accepting that the structure remains stable in a whole in which change is inevitable means giving up all benefits the accumulation of knowledge may provide. Therefore, in order to realize change in a way that it provides the most benefit, compatibility provided by internal balances is needed. And it is possible to understand this only with the entropic approach. In every branch of social sciences concerned with uncertainty, the concept of entropy is in fact a new point of view toward the world in changing balances. Then, it can be accepted that entropy is a scale for structure.

UNCERTAINTY AND COMPLEXITY

Researchers who carry out studies on complexity may be grouped into those who aim to explain the relationship between the structure and its environment (Dill, 1958; Duncan, 1972; Lawrence & Lorch, 1967; Thomson, 1967), designers who study the impacts of complexity on structure (Burns & Stalker, 1961; Galbraith, 1977), and strategists who study the impacts of structure on environment (Merry, 1999; Lindsay & Rue, 1980).

Jurkowich (1974: 381-388) puts forth four basic dimensions for defining environmental uncertainty. In this approach, the first dimension is the ability to **perceive problems-opportunities**. Within this framework, organizational reactions differ depending on whether environmental impact is perceived as a problem or an opportunity. Therefore, perception of the environment may occur as problem detection or opportunity assessment for different structures.

The second dimension of environmental uncertainty is whether the elements constituting the environment **are organized or not**. If the elements comprising the environment can be seen in an organized structure, it will be easier to perceive the impacts and forming the organizational reaction. Therefore, the environment comprising organizations is a determinative field for reaction.

According to Jurkowich, another feature of environmental uncertainty is **direct or indirect interest factors**. Multitude of factors on which interest is directly concentrated makes the environment uncertain.

In Jurkowich's definition, the fourth dimension is **complexity**. Child (1972: 3) linked complexity to heterogeneity and then action. Accordingly, the more the number of environmental factors is, the easier it is to mention complexity. Thereupon, Child emphasized that interaction of the system with different environmental factors would lead to complexity, as well.

At this point, Simon (1962: 103) put forth a different point of view, claiming that heterogeneity has little impact in terms of complexity. This view was shared by Duncan (1972), too. According to Duncan, signs of environmental complexity appear with the multitude of decision points. The more decision points form concerning environmental elements, the more the complexity increases. Simon (1962: 107) asserted that complexity could be analyzed by explaining these points and solutions could be found for problems caused by complexity through such **short steps**. Therefore, Simon claimed that complexity should in fact not be a much effective sign regarding uncertainty. Starbuck (1976: 1074) supported the opposite of this view, stressing that the actual problem about complexity is the perception of decision points. Thus, as Starbuck describes, while an environment is perceived by an organization as unpredictable, complex and variable, it can be considered by another organization "stagnant and easily understandable[2].

This view, which argues that the actual problem about complexity is the perception of decision points, stresses that symbols are the primary cognitive point in perceiving the environment, and

---

2   In organizational literature, this view is supported by imaginative dimension researchers (Peters, 1987; Pfeffer, 1981; Smircich, 1983; Morgan, 1986).

environmental complexity should be defined in this sense. Thus, organizational complexity can be defined in different dimensions according to individuals' perceptions. This kind of approach creates an approach specific to social systems, revealing more information processing processes and need to understand environmental complexity. If the social system encounters environmental factors, the problem will become clearer toward information processing (Thomson, 1967: 70). This problem about information processing can be prevented with the emergence of elements that are able to carry out information storage in the corporate scale, since individuals' abilities are limited.

One of the earliest studies for defining complexity in social systems by rate of change is the well-known Tavistock studies of Burns and Stalker (1961). The Tavistock studies are of special importance for this study in that, although they were essentially concerned with organizational structures, they used the change of environmental elements as an indicator in the modeling phase. Following the Tavistock studies, Lawrence and Lorch (1967) stated that speed of change alone could not be determinative enough for environmental uncertainty, and beyond this, the duration of obtaining feedback would also be effective for the outcomes of certainty and decisions in obtaining information. Lawrence and Lorsch asserted that speed of change, uncertainty of obtaining information and feedback duration constituted the environmental picture, and the structure shaped itself within this picture. These findings expanded the Tavistock studies in terms of understanding.

Environmental interaction started to be handled mainly from the structural aspect thanks to the studies of Emery and Trist (1965). The "dependent pattern of organizational environments", put forth by Emery and Trist, classified the environmental elements with which organizational contact could be made. Definition of these environmental elements constitutes the basis of organizational behavior. In order to make the definition in question, the speed of change of environmental elements must be put forth. In addition, the power of interaction among the elements must be known. Differently from other studies, Emery and Trist (1965: 25) emphasized with this point that environmental elements affect each other. Thus, the fact that interactions of environmental elements directly influence the organization was defined. According to the speed of change of the interaction of environmental elements and their interaction with other elements, four different types of environment could be defined. Firstly, **stagnant and dispersed environment** may be considered, in which speed of change is too low and interaction of environmental elements is weak. In this case, elements of organizational environment will be few and have a simple structure. Thus, defining the environment will require very little information.

The second type of environment, as Emery and Trist (1965: 31) suggested, is **environment with low speed of change but intense environmental interaction**. Increase of interaction among environmental elements brings about grouping. Thus, behaviors in the organization's environment will be defined with more complicated expressions and formulas. Here, the organization will have to focus on the environmental structure because the environmental structure has more power of influencing the organizational structure. The third type of environment, in which speed of change is high and relationships are intense, is **unbalanced and reactive environment**. This type of environment shapes organizations and the interaction among them. Expression of behaviors of elements constituting the environment with complicated formulas is not seen in this type of environment but the complex structure of the interaction among elements increases environmental uncertainty.

An environment with complex interaction, in which environmental variables are expressed with complex formulas, is **turbulent environment**. According to Emery and Trist, this is the most complex type of environment (1965: 29). Emery and Trist's interpretation of the interaction between the interior and the environment of the organization is shown in Figure 2. Emery and Trist's definition of environment through the change and interaction aspects of the elements within it created the basic knowledge that environment is a dynamic setting. Emery and Trist (1965: 30) stress that, in a turbulent environment, interaction among variables influence the organizational structure at the lowest level. Emery and Trist, in a sense, examined the organizational environment by observing it from inside the organization with this approach. Perception of the interaction between the organization and its environment as only a relationship from the organization toward the environment has caused the environment's impact on the organizational structure not to be clearly understood. This unidirectional perception has brought about the fallacy that the change of environmental fabric influences the organizational structure less than other reasons.

According to this view, the elements within the environment create the change of the environmental fabric. Change of the environmental fabric will also lead to the change of Emery and Trist's "first-level environmental factors" (1965: 24). Thus, change of all organizational reactions and structure, primarily the reaction of establishing relationship with the environment from inside the system (cooptation), will take place. This interpretation can be more clearly defined with the acceleration of innovations' emergence, which was exemplified by Alvin Toffler with regard to speed of change (Toffler & Toffler, 1995: 47-48).

It can be considered that interaction structures of systemic variables are also an important factor in understanding environmental complexity. Emery and Trist (1965: 24) mentioned von Bertalanffy's (1950) equality emphasizing the input-output balance, stressing the importance **negentropic** (creation of order by reverse occurrence of entropy, which is the criterion of the structure) structures (1965: 30).

Systems process the materials they obtain from the environment and present them back to the environment as outputs. Integrity cannot be maintained with a structure in which the energy from inputs is equal to the energy it creates in outputs. In order to maintain continuity and preserve the structure, the input energy must be greater. This margin is required for preserving integrity and using it during the process. This approach constitutes the building block of the environmental theory in that it emphasizes the importance and dependence of processes.

Terrybery (1968: 590-601) stressed the importance of communication in change, underlining that Emery and Trist's four environments are a development model and organizations are exposed to the impacts of other factors by losing their independence within this development with an increasing speed.

Emery and Trist's conceptual presentation of images within the environment's dependent fabric can also be seen in J. Thomson's studies (1967). The interaction of environmental variable processes among themselves put forth by Emery and Trist is not emphasized in this analysis. And Thomson's definition handles environment with its simple-heterogenous and stagnant-variable dimensions (1967: 38-47).

The interdependency among organizational characteristics is genericaly shown in Figure 1. This type definition shows resemblance to the different types of definitions of Pfeffer and Salancik (1978), Aldrich (1979), Jurkowich (1974), Mintzberg (1979) and Scott (1981). For example Aldrich defined environmental dimensions as **openness** (capacity), **variability** (stagnant, dynamic, turbulent), and **complexity** (heterogeneity–homogeneity).

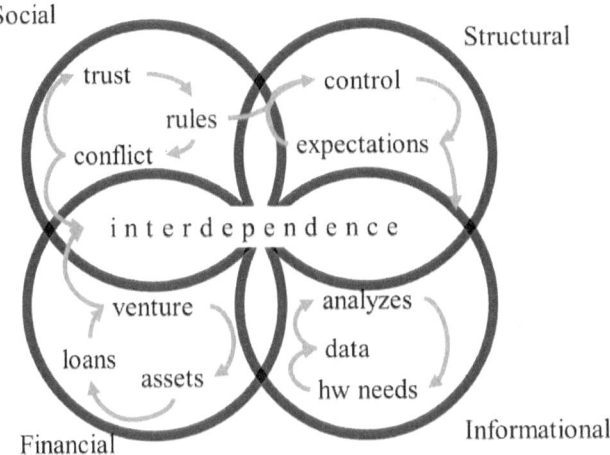

Figure 1: Organizational Interdependency

And the relationships among these dimensions are variable and structural ones. Pfeffer and Salancik state that concentration and openness affect conflict negatively. Conflict and disconnectedness are treated as social mobilizers. These two mobilizers consequently lead to complexity. Alongside this, Miles and Snow (1978) explain Aldrich's environmental variability with the unpredictability of change instead of the speed of change in the environment. Thus, a new dimension is added to the definition of change.

The fact that change of environmental factors cannot be precisely predicted also prevents reaction to be shaped in this regard. Besides this difficulty of prediction, that change does not follow a linear slope play an important role in this respect, as well. Change of any of the environmental factors directly affects the environmental structure. In this case, the phenomenon described as turbulent environment may become chaotic. Thus, a complete uncertainty emerges regarding not only the dimensions of the change but also the start of the change.

These tides created between the regular environment and the turbulent environment depending on the change of environmental factors causes the change to bifurcate. If, at first, environmental change is defined with a curve dependent on the change of the elements within the environment, this curve will begin to bifurcate due to the impacts of the elements' nonlinear behaviors. At this point, it is possible to make a definition with a nonlinear mathematical expression by drawing upon the fact that environment has a variable structure with its visible aspect and is difficult to predict. This definition is the expression of the dynamic movement of the nonlinear behavior, rather than a picture containing environmental expressions.

Within this framework, it can be examined by considering the expansion (growth) of the environment. If the fact that environmental magnitude has a limit is taken into consideration, the value for reaching this number can be called S. In order to prevent the term S to grow infinitely, it must be multiplied by its mathematical opposite *(1 - S)*. Changing of the S value between 0 and 1 indicates the expansion or contraction of the environment by percentage. Thus, if the mathematical definition is reconsidered, an equality like

$$S_{final} = S_{previous} \, (1\text{-}S_{previous})$$

is reached. In order to emphasize the difficulty of predicting the behaviors of elements within the environment, a control variable (r) may be added and the mathematical expression of the nonlinear environment may be turned into

$$S_{n+1} = r \, S_n \, (1\text{-}S_n)$$

Considering that the amplitude of the environment in the beginning is any magnitude such as 0.1, and in a setting where the r constant is 3 (i.e., in the case that it has a threefold impact on the change of environmental factors).

Considering that the change constant (r) is less than 1 and the initial expansion value is different, the quality of the change will naturally differentiate, as well. In this case, it can be expected that there will be a different figure for each value. In the cases that the change constant is found to be greater than 3 as a result of different trials with different values, the environment can be seen to repeat itself (Figure 2). In this case, the environment bifurcates by developing in two different directions. Feigenbaum (1978: 35) observed that this bifurcation value's rate of repeating itself (i.e., the ratio of the length between both bifurcations to each other) is constant and this constant is approximately 4.669... If the environmental conditions, which Emery and Trist (1965) called the turbulent environment, are worked until infinity through mathematical expression, the ratio of the change lengths to each other will equal the Feigenbaum number. The expression of the environment's dynamic structure can be adapted to real life only if the mathematical expression of all relationships and characteristics in the environment can be achieved.

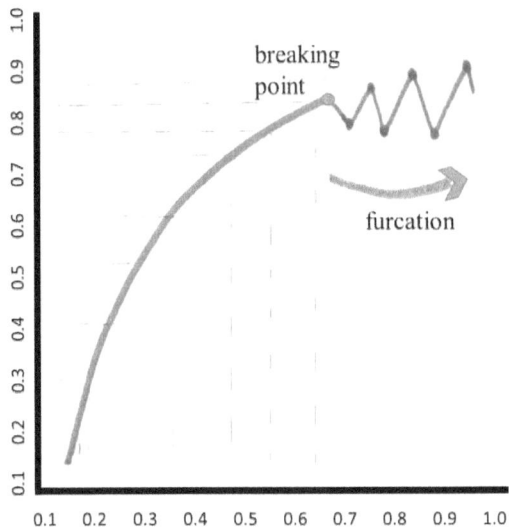

Figure 2: Bifurcation Slope of Organizational Reaction

Then, the definition of environmental conditions is carried out in different ways depending on the abilities of the perception organs. Within the system, these perception mechanisms can be interpreted as boundary elements. Thus, the reaction mechanism is structured through a boundary to be formed between the system and its environment.

Boundary can also be perceived as a shell that helps the structure to be preserved. In cases that the structure cannot be preserved, that is, in entropic settings where dissipation takes place, boundary elements will lose their ability to determine behaviors. Such a situation, which can mean that the shell has been broken, will cause entropy to increase and dissipation to occur.

Systems perform complex adaptation behaviors as simple commands and with the help of mobilizers. Mobilizers make the mobilization continuous and provide improvement. The main duty of mobilizers is to constantly scan other mobilizers in the environment and the system, and perceive extraordinary behaviors and changes. If, during this perception process, a factor to react to is encountered, a reaction is developed within its internal model or with the help of schemes.

NONLINEAR DYNAMIC SYSTEMS

Systems that behave linearly provide a proportion between inputs and outputs. These behaviors can be demonstrated with a plain graphic. For example, if a weight of 1 kilogram is hung to the end of a solid rubber band, the band stretches ten centimeters; if a weight of 2 kilograms is hung, then the band stretches twofold, that is, twenty centimeters. Since there is a proportion between inputs and outputs, the system's behavior can be predicted and thus be controlled generally. Linear systems have an orderly and predictable style and change continuously, predictably and controllably.

Linear impacts are not frequently seen in real life. For instance, if ten aspirins are taken, they do not necessarily relieve the headache tenfold. In nonlinear systems, there is not a linear proportion between input and output or cause and effect. Contrary to nonlinear causality, linear causality acknowledges unidirectional interaction and denies the interactive bidirectional causal relationship between objects. The faster a sled goes, the more its friction with snow will be and this affects the sled's speed. In these types of mutual causalities, there is usually an exponential relationship instead of a proportion between the input and the output. As can be seen in the event where Nicholas Leeson shook up the 200-year-old Barings Bank and global stock markets[3],

---

3   In 1995, Nicholas Leeson caused the bank where he worked as a broker to overborrow and go bankrupt due to the wrong decisions he had taken. The fact that the bank went bankrupt, although he was not a part of the senior management's decision-making body and there were many other employees like him, shows that small impacts seeming insignificant can create big outcomes.

a small input can result in a big output. This resembles the triggering of an avalanche by a snowball. Complex systems do not demonstrate orderliness, accuracy, and the ability of constantly changing and reversion. These characteristics are represented linear, isolated systems inherited from Isaac Newton, Francis Bacon and René Descartes. In complex systems, predictability is limited and so is the ability of planning and controlling events.

The expression of **nonlinear** is about understanding mathematical models that explain systems. Until the interest in systems multiplied, most models had been analyzed as if they were linear systems. Due to this approach, when mathematical formulas representing systems' behaviors are graphed, the results appear as a straight line. Algebra is a mathematical method Newton used for demonstrating the change in systems with straight lines. However, statistics is a process that changes nonlinear data for making it linear.

Complex systems may demonstrate complex behaviors with simple inputs. An association cannot be formed between the quality of the phenomenon creating the behavior and the complexity of the behavior. This even applies to simple equations. Considering that, for a company, the increase in marketing brings about profit growth at certain rates, this relationship can be inferred to be linear:

$$G_f = a \, X_f$$

Here, G is the profit gained during the term f,
X is the R&D expenses during the term f, and a is the fixed rate.

Thus, it can be thought that R&D expenses create profit growth at a certain rate (a) after all other changes are ignored and fixed in order to achieve simplicity. Considering that this enterprise allocates a certain amount of its end-of-period profit to next period's R&D expenses, the formula of

$$X_f = b \, G_{f-1}$$

can be achieved. This equation expresses the relationship between R&D expenses and previous-period profit. When these two equations are combined, the formula of

$$G_f = a \, b \, G_{f-1}$$

is achieved. Thus, a relationship expressed with the decision variable is found. When the *a.b* expression is represented with the *s* constant to ensure simplicity, the equation of

$$G_f = s \, G_{f-1}$$

is achieved. The graphical expression of such an equation will be as shown in Figure 3. Depending on the changes in the s constant, it is possible to achieve lines with different slopes. If the s constant is chosen to be less than 1, the profit *Af-1* in the previous period and the profit *Af* in the present period form the *s<1* slope. It is a linear form.

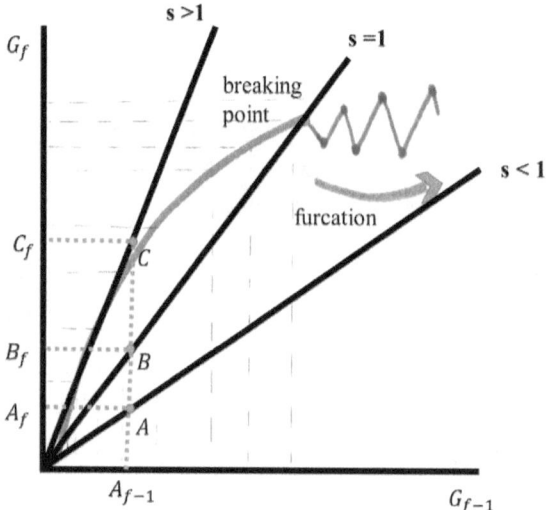

Figure 3: Linear vs. Nonlinear Behavior Slopes

If s is zero, and the middle slope is greater than 1 with the same values, then the slope on the top is achieved. As it happens in the real world, the return of advertising expenses will decrease after a saturation point. In this case, profit will increase to a certain level and then start to decrease. In any profit *A*, in order to learn the next period's profit, a line parallel to the G_f axis can be drawn and the value it corresponds to can be checked. In this case, the next-period profit will be *B*. To find the next profit after *B*, the same process must be repeated by finding the expression of the Bf value on the Gf axis. For this, a 45-degree slope will be created. Since this slope is achieved by taking the same value on both axes, the value on the *Gf-1* axis can be achieved by looking at the B value's projection on this slope. In the figure, the Bf point is the expression of this situation. And a parallel to be drawn on the Gf axis from the Af point will show the next profit point with the intersection point on the slope. In the figure, this point is the C point.

When these steps are repeated, an orientation can be seen toward the point where the 45-degree slope and the profit slope intersect. This cycle expresses a tendency toward profit-advertising expenses that diminishes by itself, depending on the profit rate decreasing in the previous period. Thus, regular profit flow can become possible, as well. This behavior of the enterprise, which develops depending on the profit and advertising expenses, is shown in Figure 4.

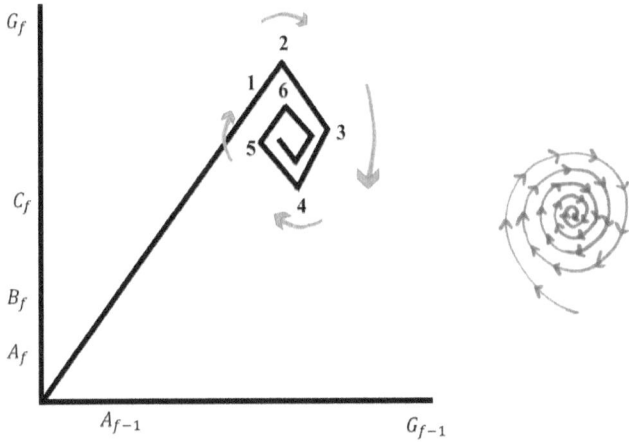

Figure 4: Combining at the Zero Point

This means, in a sense, that the system repeat itself. The system thus orients itself toward the intersection point by itself depending on the constant value. As such, the **self-organization ability of dynamic systems**, which is one of the foundations of chaotic behavior, is expressed.

While the prevailing behavior for the enterprise in question here oriented the system toward a certain stationarity, the slope could have been as shown in Figure 5 depending on the s constant having a higher value. In that case, the enterprise could have been restricted by the attraction of chaotic behavior by remaining within the boundaries of the *1234* quadrilateral. After escaping from the attraction of chaotic behavior, the system will again come under the influence of the same attraction and continuously go through tides between order and disorder. These tides will keep happening at fixed intervals in the bifurcation curve but with a slope orienting toward the intersection point.

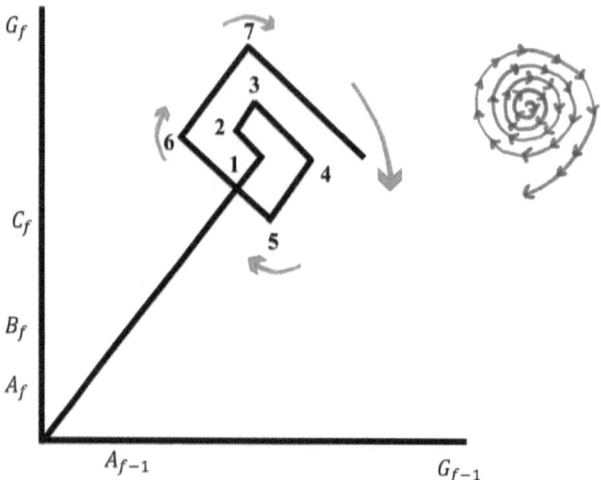

Figure 5: Orientation toward Chaotic Behavior

In this case, one more criterion needs to be added for the enterprise, which anticipates profit growths related to advertising expenses. The margin of error, which forms due to the fact that a considerable number of criteria impossible to control in real life influences the systems, will make the equation more realistic. By including the margin of error in the formula with a probability, the equation of

$$G_f = s \ G_{f\text{-}1} \ (1\text{-}G_{f\text{-}1}) + h$$

is achieved. Here, h denotes the error value and the $(1\text{-}G_{f\text{-}1})$ expression denotes its probability. By adding the error value to the constant in the formula, a simpler equation of

$$G_f = s \ G_{f\text{-}1} \ (1\text{-} \ G_{f\text{-}1})$$

is achieved. Thus, the enterprise is able to formulate its behavior by taking the risk of encountering errors with certain probabilities. Table 1 demonstrates the probable results of this equation with different s constants. Thus, as in the bifurcation slope, the areas where behaviors are influenced by chaotic attractors and escape from this influence are marked.

Based on the value of s in the formula $Gf = s\ Gf\text{-}1\ (1\text{-}Gf\text{-}1)$ the behavior of the system vibrates (Goldfain, 2008; Sarmah & Paul, 2010). For example system's attractor will be proper value if s equals to any value between 0 − 3; it will be 2-period circle if s equals a value between 3 - 3.5; it will be 3-period circle if s equals a value 3.835; and it wil be chaotic if s equals to values 3.58, 3.80, or 3.85.

This situation shows the probable outcomes of simple but nonlinear behavior for the enterprise giving advertisements. Besides, the enterprise gains the opportunity to understand the chaotic environment's own order. The view that chaotic environments have a natural order within themselves but it requires effort for understanding this is marked by scientists lead by Feigenbaum. Such an environment is a novelty that has no place in conventional thinking and decision processes. Decisions made conventionally do not mean anything else than creating the linear continuation of data from the previous period into the next period. Today's information technologies can be defined as a continuation of this view. Within this framework, conventional information systems can be claimed to have a linear trend.

Simply, in linear systems a problem is divided into many small parts, then the separate solutions are added to each other and a solution is found for the whole problem. But two solutions of a nonlinear equation cannot be added to each other for another solution. Therefore, it is needed to consider a nonlinear problem a whole. That is, no one cannot, at least explicitly, divide the problem into smaller parts and add their solutions to each other. Hence, it is not surprising that there is no general analytical approach in the solution of typical nonlinear equations.

For example, water's movement while it flows through any pipe at a low speed is a property of linear behavior; it is normal, predictable and definable. But when the speed exceeds a critical value, the movement turns into turbulence movement with local vortices moving in a complex, abnormal, disorderly way like nonlinear behaviors. In this case, it can be thought that nonlinear systems differ from linear systems as explained below:

Firstly, the movement itself is different. Linear systems demonstrate movements that can be explained typically with soft and usual functions in space depending on time. However, nonlinear systems mostly can transition from soft movements to complex and sometimes even random movements.

Secondly, linear systems' reaction to small changes in their variables and external stimuli is mostly soft and directly proportionate to the stimuli. But in nonlinear systems, a small change in variables may lead to big changes in the movement.

Thirdly, in linear systems, a local **lump** or impact normally spread and disappear in time. Contrarily, in nonlinear systems, there may be highly compatible and balanced local structures. These endure for a long time like the vortices in the turbulence flow, or forever as in some ideal mathematical structures. The incredible order reflected by these steady and compatible structures contradicts the complex and disorderly movement they are in. This behavior of linear systems resembles the nine women giving birth to a child in one month. In order to explore the answers like turbulence, vortex and disorder given to only the small changes in necessities, countless linear systems have been planned as the resultant of soft-flowing sub-projects under the system cascade.

These distinctive features of nonlinear systems are usually met with a great uncertainty, and a reaction is produced from the efforts renewed by remaining dependent on the processes and structures failing at first in order to force the system to a linear flow with ultimate sensitivity.

The **divide and rule** linear thinking process assumes that a system can be divided into parts and each part can be programmed to be pieced together again for creating a whole. But this does not apply to nonlinear systems; that is, the divided parts are not necessarily able to form a whole.

At this point, the easiest method mostly preferred is hierarchical decomposition. In this approach, which implementers of organizations are familiar with, organizational integrity is broken up through segmentation and energizing. In fact, the segments perform an automation function. Therefore, the whole organization must have a single viewpoint. Such a point of view typically resembles the behavior of a department director whose function is automated by a computer. Since all departments in an organization are automated, the viewpoints programmed based on the needs of each department are seen as a discharge system, just like a **stove chimney**. Eventually, an initiative is established to achieve and **integrated** system by somehow piecing divergent viewpoints together. This usually makes the problem more complicated instead of solving it. Just like a building's chimney system, black smoke fills all rooms the moment when the compatibility of the viewpoints is lost. Thus, in such approaches, it is usually automation that is meant when mentioning compatibility.

However, what is actually meant when compatibility is uttered in a social CAS is different synchronization-based behaviors. Since the structures of their CASs are based on differentiation, automation cannot be mentioned for the compatibility of behaviors at the starting point. The unpredictability inherent in their CASs prevents this. This results from the fact that behaviors irreversibly emerge by themselves from the beginning.

CAUSALITY AND DEPENDENCY ON STARTING POINT

The quality of the mobilizer starting the behavior also plays an important role, alongside the outcome of behaviors, in understanding the behaviors of dynamic systems. Mobilizer is the element that consciously or unconsciously starts the radical change in the whole system. While an entrepreneur, competitor, customer or any personnel may be the mobilizer for enterprises, for natural environments the mobilizer may be any event such as the hunting of an animal, withering or thriving of plants. The motion started by the mobilizer within the system can spread over the whole system, causing substantial behavior changes.

Considering that systems' interaction with the environment is a function of behavior, system behavior is very difficult to predict, almost impossible to formulate, and quite complex. Just as twin siblings who have grown up in the same environmental setting and exposed to the same influences can have completely different or at least non-identical characters, systemic behaviors are in fact entirely unique. This uniqueness is created according to the quality of the mobilizers and the characteristics of the system, and with the influence of many other factors. Increasing or decreasing the number of these factors are naturally possible. Every system is influenced by many phenomena in accordance with their own characteristics. The question of which factor affects the system behavior and how much it does so, can only be answered with special modeling approaches. This subject is further analyzed in the following chapters of the book.

The factor affecting the system behavior, whether it be from the external or the internal environment, creates a peculiar impact. Every system is equipped with different features in terms of perceptive competence. Systemic perception is created by the sub-systems they contain. Perception ability of the system can be considered a function of the sub-systems creating the perception and the perceivers' characteristics.

Every system also demonstrates differences in terms of interpreting what it perceives, as well as the ability of perception. In fact, this also results from the different interpretation abilities of sub-systems or system parts. From the same point forward, it can be thought that the reflection of knowledge and performance of activities also bear a characteristic belonging to the system, completely in accordance with this composition.

At this point, it can be thought that systems are unique entities, regardless of how much structural similarity they share with other systems. What creates this uniqueness lies at the foundation of their entities. Organizations are organisms that are constituted by not just the labor force of their employees but also their emotions and intellectual abilities. These organisms constantly change from the moment of their creation on in accordance with the changes in the values they embody. The change of the organizational structure and then behavior is based on these differentiations which are not quite clear in the beginning. The differentiations in question can develop and become significant in time.

This situation can be visualized with the dropping of a pencil with sharp tip. Considering that a pencil with a sharpened tip is being hold with one finger on a table, it cannot be predicted at first on which side it will drop. If the pencil's tip is sharp enough and the finger holds it upright enough, it is almost impossible to predict the side it will drop on. When the finger is withdrawn, the pencil starts to bend toward any side. At this point, it has become possible to predict. At first, the pencil bends toward a certain direction with a relatively slow motion. At this point, if the pencil's movement can be frozen, it will be possible to predict on where the pencil will drop with a certain margin of error. When the pencil resumes its movement, it will speed up a little more and at this new point it will make more accurate predictions possible. When the pencil is observed at a point just before it drops, firstly it will be almost precisely known where it will drop, and secondly it will have reached its highest speed. If research is done into the reasons that cause the pencil to be in that very situation, many different factors can be found: A strike on the pencil while withdrawing the finger, a nearly insensible breeze through the table, or an unnoticeable mistake caused by pencil sharpener may have started this mobilization. No matter what the reason is, this mobilization is unnoticeably insignificant and small in the beginning. The mobilization will increasingly gain more importance while the pencil drops. It will even be thought that the actual reason creating the dropping point of the pencil is the action of this mobilizer.

If the pencil is let go a few more times in exactly the same way, it will drop on different points due to different impacts but with the same structural dynamism. It is possible to observe the same dynamics also in the organizational structure: An organization starts to determine its behavior from the first phase of foundation forward with the impact of many factors. The behavior is only shaped in the final phase with noticeable impacts.

For example, the existence of the organization starts with the entrepreneur's dream. From this point forward, it constantly changes, develops and becomes more original. As a new employee is recruited, as an employee quits, as new competitors show up, as the number of shareholders increases or decreases, the organization always changes. While some of these changes are visible and concrete, others become noticeable only after a long time.

This situation expresses a somehow indistinct bond to the starting point. This bond is not quite distinct because it can change any moment until the conclusion point is reached. In addition, it only becomes possible to clearly reveal this bond in the last phases. This relationship of organizational behavior with the starting point can only be understood after its signs appear. The chaotic system discovered by Edward Lorenz is a nice example that explains this situation (Gleick, 1987: 23). In the example, a waterwheel is examined in order to achieve simplicity. The waterwheel is a simple mechanism which has buckets hung around it and turns as the bucket at the top is filled with water (Figure 6). The important point here is that the wheel will not turn unless the bucket at the top is completely filled. In addition, if the bucket is filled slower than required, it will not be heavy enough to beat the friction force of the wheel and consequently the wheel will not turn. Then, the water must flow faster than a certain speed.

If the water's flow is fast, the top bucket's weight will mobilize the wheel. As the flow gets faster, the turning movement will become chaotic due to nonlinear behavior characteristics within the system. How much water is filled in the bucket while they pass under the flowing water will be determined by the rotation's speed. If the wheel's rotation is too fast, there will not be enough time for the buckets to be filled. Besides, the buckets will begin to go up again without being able to discharge the water inside them.

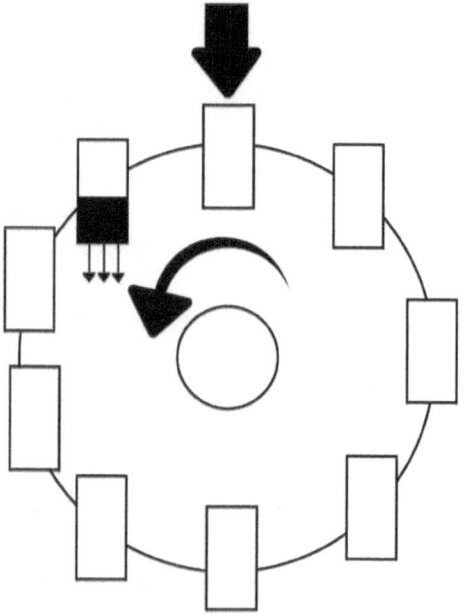

Figure 6: Lorenz's Waterwheel
Source: Adapted from Gleick (1987: 23)

As a result, the heavy buckets on the upward side will cause the rotational movement to slow down and then the wheel to rotate in the reverse direction (Figure 7). Indeed, Lorenz discovered that the rotational movement would frequently reverse in long periods and never continue with a regular pace, and in the meantime never occur in a predictable way. As a result of such a movement, behaviors that never intersect, as demonstrated in Figure 7, and may develop in the reverse direction in time. This graphic, which Lorenz likened to butterfly wings, is one of the first expressions of chaotic behavior.

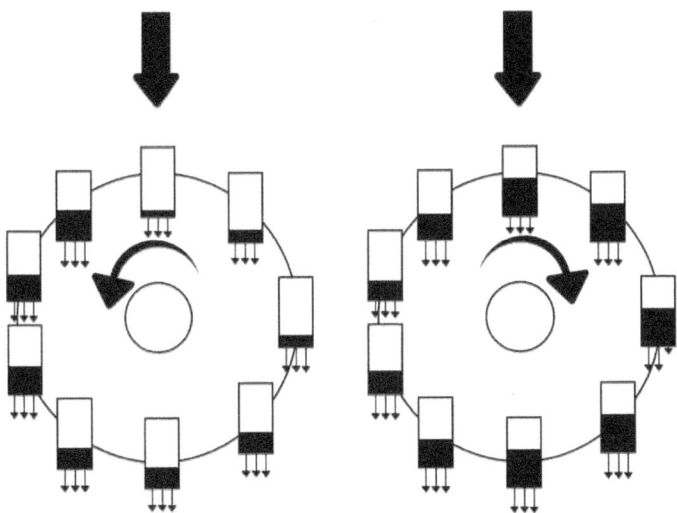

Figure 7: Wheel Rotating in Reverse Directions
Source: Adapted from Gleick (1987: 23)

When the butterfly effect is taken into consideration, characteristics of chaotic behavior that are never the same and predictable comes to the forefront. The graphic of the butterfly effect can be started from any point and the result will be the same. Besides, it can be seen that two points very close to each other repeat the same movement forever without ever intersecting each other or themselves.

No matter how close they are to each other, the reason why two different points demonstrate the same movement but in a different way is free dependence on the starting point and self-similarity. The reason why the same movement is observed in both behavior patterns is dependence on the starting point. And differentiation of behaviors from each other expresses self-similarity. The fact that human genes demonstrate different characteristics for everyone although they are made up of four bases is an example of self-similarity. The fact that all genetic codifications are within the same behavioral structure while there is a unique codification for everyone indicates free dependence on the starting point. These behavior patterns are among the basic features of complex systems. Therefore, it is possible to see these behaviors not only in biological systems but also social systems. If the starting point in social systems is defined, it is probable to think that the behavior's general structure will emerge dependently on this. At this point, it will become possible to utilize an important tool in demonstrating the chaotic structure of social systems' behaviors (Figure 8).

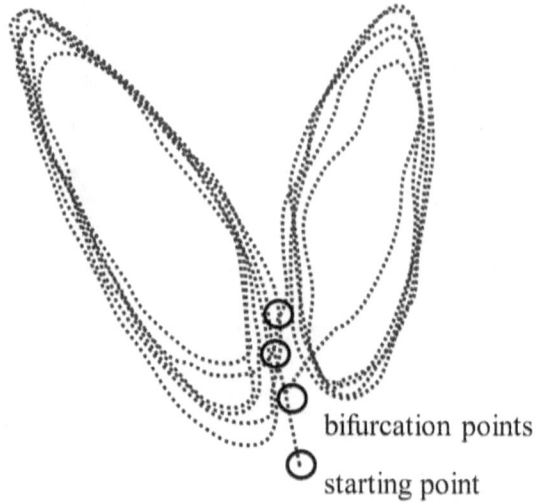

bifurcation points

starting point

Figure 8: Generic Representation of Butterfly Effect

Edward Lorenz (1998: 20), while modeling meteorological dynamics in his office, observed that the model made great errors after certain duration of prediction. When he investigated the reasons of this, he found that the error in the model happened due to the rounding up of the permillage of numbers after the fraction. In the model, the permille error caused it to snow in July.

In social systems, it cannot be possible within the dynamism of daily life to predict the impacts of many errors on policies and culture within this framework. Within the social system, these oscillations emerging due to various reasons, such as differences of perception capacity, environmental impacts and so on, prevents the system's behavior to be predictable and assimilable.

At this point, it can be thought that behaviors and dependently environments of systems are under chaotic behavioral patterns. In the light of this thought, it is possible to see that behaviors can be distinguished due to their self-similarity but predictions cannot ever be made. Here, the only point where systemic reaction can take place will be the learning of the behavior's dynamics which is carried out by developing the perception ability. When the fact that, in real life, weather forecasts can be made only for very short periods of time despite today's technological capabilities is taken into consideration, it can be seen that the system-environment interaction also demonstrates an unpredictable structure.

This self-similar and nonlinear structure the interaction between the system and its environment is shaped through organizations' characteristics of perception, interpretation and reacting. Then, systems' ability to perceive, interpret their perceptions distinctively and make them continuous becomes especially important for analyzing behaviors. These characteristics, which can be regarded as the structural characteristics of systems, are peculiar to each system. Therefore, just as it is observed in individual behaviors, different systems can give different reactions to the same impacts. These differences can be the basic determinants of system behaviors, no matter how much unnoticeable or insignificant they are at the start. This is of utmost importance in revealing the system behavior.

The three characteristics of systems (holism, dynamism and causality) requires of special logical consistent models to be developed for understanding systems. It is clear that intellectual models that are not consistent with these three characteristics will also be limited in terms of the ability to explain the behaviors of systems. Therefore, the necessity to develop an approach based on non-discrete dynamic variables for understanding systems increases every day.

Fuzzy Logic

After Aristotle formulated the **law of excluded middle** (1995), the legacy of Aristotle had not questioned till Russell. Russell (2009:102-103) asserted that "Class may be defined either extensionally or intentionally. That is to say, we may define the kind of object which is a class, or the kind of concept which denotes a class". A brilliant question is raised in this perspective; how, in Aristotelian approach, can a set of all sets, which are not members of themselves, be a member of itself? This inquiry is known as Russell's paradox. After Russell, Lukasiewicz, in the same perspective, improved the 0 and 1 values in the Aristotle's Logic and asserted the expression of [0,1,2] by adding a third value (1957). After Lukasiewicz, Donald E. Knuth (1968) suggested that the integer values of [-1,0,1] be used instead of Lukasiewicz's value of [0,1,2]. The studies in question did not gain recognition at that time [1,0].

From the systems thinking perspective, a system is a set of systems. To understand the behaviors of the system clearly, the members and definitions used among the all systems parts should be typical including the whole system. Since systems thinking has been developed based on the understanding that systems are organic structures sensitive to the initial conditions and external environment, the logical system it requires must also demonstrate this understanding. So, members and definitions should be part of the different systems at the same time with different amount of membership percentages.

Today, the logical system that can fulfill such a need is Fuzzy Logic. Fuzzy Logic is a logical system that is grounded by being inspired by the human brain. Instead of the values of 0 and 1 within Aristotle's Logic, it makes it possible to use other numerical values between these values. Hence, it helps verbal concepts like less, more, some to be digitized, and reduces information loss. In Fuzzy Logic, in which degrees of membership between 1 and 0 are used, an element can belong to both sets. Within this scope, it is provided that both sets contain elements in a certain ratio by determining the degrees of membership for each set. It handles uncertain, time-varying, complex systems similarly to human thinking. The founder of fuzy logic, Omer Lutfu Aliasker Zadeh defined the features of fuzzy logic which are not possessed by traditional logical systems as follows (1990:101-102):

1. In fuzzy logic the truth values are allowed to be fuzzy sets labeled true, quite true, very true, more or less true, mostly true, etc.

2. In fuzzy logic, the quantifiers such as most, many, few, several, usually, etc. are interpreted as fuzzy numbers which serve to describe the absolute or relative cardinalities of fuzzy sets.

3. From the frequentist point of view, fuzzy quantifiers bear a close relation to fuzzy probabilities, e.g., likely, not very likely, unlikely, etc., which underlie much of the commonsense reasoning which we employ in our daily decision making. Through the connection between fuzzy quantifiers and fuzzy probabilities, fuzzy logic provides a machinery for qualitative decision analysis in which fuzzy utilities are computed from fuzzy probabilities and fuzzy payoff functions.

4. Fuzzy logic provides a mechanism for dealing with the hedges (very, more or less, quite, somewhat, extremely, etc.) by interpreting them as operators which act on fuzzy predicates.

Zadeh published many scientific papers regarding fuzzy logic and as the founder of fuzzy logic, influenced the thinkers and academicians of his time (Spagnolo, 2003). In Zadeh's definition (Zadeh & Desoer, 1963: 65) a system "S is a partially interconnected set of abstract objects G1, G2, G3. . . termed the components of S. The components of S may be oriented or non-oriented; they may be finite or infinite in number; and each of them may be associated with a finite or infinite number of terminal variables". Zadeh wrote the article titled "Fuzzy Sets" in 1965, in which he defined that a fuzzy set is a class of objects with a continuum of grades of membership (338). Fuzzy Logic was first integrated into the control system of the steam engine by Mamdani and Assilian (1975). After this starting point the use of fuzzy logic is spreading as a new approach to system analysis. Zadeh explains his new approach as a "departure from the conventional quantitative techniques of system analysis" and listed the main distinguishing features of this new technique as follows (1990: 96):

1. Use of so-called "linguistic" variables in place of or in addition to numerical variables,
2. Characterization of simple relations between variables by fuzzy conditional statements,
3. Characterization of complex relations by fuzzy algorithms.

According to Zadeh (1990: 98-100) the expression of thinking through direct numerical values is not an appropriate method for the human mind. Zadeh (1975a) emphasizes that the real world contains thousands of values and similarities between the 0 and 1 values. Within this scope, Fuzzy Logic offers the opportunity to carry out operations with imprecise values. In his studies, Zadeh grounded these opportunities on certain disciplines and principles (1975b). Fuzzy Logic contains rules even though the variables are expressed with fuzzy numbers.

The structure of the world, which is complex, incomprehensible through linear thinking and free dependent on the starting point, has always been a problem for strategy developers who cannot change their intellectual models. Intellectual models are required to be restructured. Seeking answers to the questions of how everything as a whole, each entity affects those around them and where this interaction stems from constitutes the basis of this new intellectual model.

According to Aristotle, who played the biggest role in the establishment of the logical system in the world, something could be either hot or cold. There is a very sharp line between a hot object and a cold object. Let this line be 30 degrees. At 29 degrees an object must be called cold, while the object at 31 degrees must be hot. However, according to the viewpoint of Zadeh – an Azerbaijani Turk – in the 1960s, this situation expresses fuzziness. Then, fuzziness prefers using partial expressions in defining entities; there is warm between hot and cold. Accordingly, for instance, an object at 31 degrees is 85% hot and 15% warm (Figure 9).

At this point to which the evolution of scientific thinking has brought humanity, being able to see the whole, define the relationships among parts and monitoring the impact has become the key to **correct thinking**. This thinking in logic is based on defining the universe as a system and improving the understanding within this definition. And expressing the world this way as a whole refers to a different paradigm.

In daily life, too, expressing entities within a whole and seeing them as a system are like defining them fuzzily. Then, operational rules of systems can also be shaped within the understanding of Fuzzy Logic. For example, a climate system has two entities; controlling the condition (cold, warm, normal, hot, very hot) and controlling the speed of engine (stop, slow, normal, fast, and very fast). Then an operational rule is **if it is hot speed should be fast**.

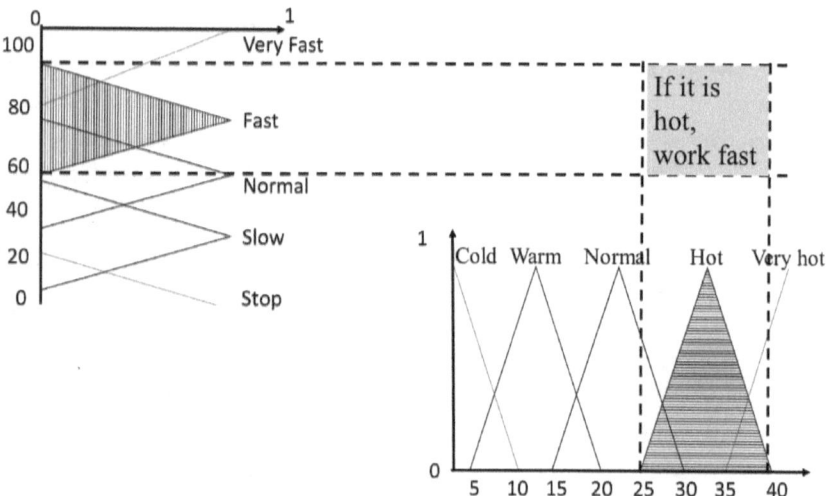

Figure 9: Fuzziness in Thermostat

In the example of thermostat, how the system will behave can be explained via Fuzzy Logic most sensitively and environmentally-consciously. Then, the air conditioning system will be enabled to operate fast between 25 and 37 degrees (Figure 9).

In conclusion, at the basis of the approach called systems thinking lie the concepts of holism, dynamism and causality. Developing intellectual approaches within the frame of these concepts is a basic requirement for using systems thinking. And the logical system that supports the usage of these concepts is Fuzzy Logic.

REFERENCES

Aldrich, E. H. (1979). *Organizations and environments*. N.J: Prentice-Hall.

Aristotle IV, B. C. E. (1995). Rhetoric for Alexander. In J. Barnes (Ed.), *Complete works of Aristotle*. Princeton, NJ: Princeton University Press

Boltzman, L. (1995). *Lectures on gas theory* (Stephen G. Brush, Trans.). MA: Dover.

Brandstädter, K., Harms, U., & Grossschedl, J. (2012). Assessing system thinking through different concept-mapping practices. *International Journal of Science Education, 34*(14), 2147-2170.

Burns, T., & Stalker, G. M. (1961). *The management of innovation*. London: Tavistock.

Child, J. (1972). Organizational structure, environmental performance, the role of strategic choice. *Sociology, 6*, 1-22.

Davidz, H. L., & Nightingale, D. J. (2008). Enabling systems thinking to accelerate the development of senior systems engineers. *Systems Engineering. 11*(1), 1-14.

Davis, A. C., & Stroink, M. L. (2016). The relationship between systems thinking and the new ecological paradigm. *Systems Research and Behavioral Science, 33*(4), 575-586.

Dill, W. R. (1958). Environment as an influence on managerial autonomy. *Administrative Science Quarterly*, 409-443.

Duncan, R. B. (1972). Characteristics of organizational environments and perceived environmental uncertainty. *Administrative Science Quarterly*, 313-327.

Emery, F. E., & Trist E. (1965). The casual texture of organizational environments. *Human Relations*, 18, 21-31.

Esen, H. Ö. (1985). *İşletme yönetiminde sistem yaklaşımı*. İstanbul: İstanbul Üniversitesi Basımevi.

Feigenbaum, M. (1978). Quantitative universality for a class of nonlinear transformations. *Journal of Statistical Physics*, (19), 25-52.

Galbraith, J. R. (1977). *Organization design*. Addison Wesley Publishing Company.

Goldfain, E. (2008). Feigenbaum attractor and the generation structure of particle physics. *International Journal of Bifurcation and Chaos*, *18*(03), 891-896.

Gleick, J. (1987). *Chaos: Making a new science*. NY: Viking Penguin.

Hershey, D. (1991). A symmetry number for the analyses of organizational structure. *Systems Research, 8*(2), 101-102.

Hung, W. (2008). Enhancing systems- thinking skills with modeling. *British Journal of Educational Technology, 39*(6), 1099-1120.

Jurkovich, R. (1974). A Case typology of organizational environments. *Administrative Science Quarterly, 19*, 380-394.

Knuth, D. E. (1968). Semantics of context-free languages. *Mathematical Systems Theory, 2*(2), 127-145.

Lawrence, P. R., & Lorsch, J. W. (1967). *Organization and environment*. Homewood, IL: Richard D. Irwin.

Lindsay, W. M., & Rue, L. W. (1980). Impact of the organization environment on the long-range planning process: A contingency view. *Academy of Management Journal, 23*(3), 385-404.

Lukasiewicz, J. 1957. *Aristotle's syllogistic.* 2nd Ed. Oxford: Clarendon Press

Mamdani, A. S., & Assilian, S. (1975). An experiment in linguistic synthesis with a fuzzy logic controller. *International Journal of Man Machine Studies, 7*, 1-13.

Maxwell, J.C. (1871). *Theory of heat* (Reprint 1968). London: Longmans.

Merry, U. (1999). Organizational strategy on different landscapes: A new science approach. *Systemic Practice and Action Research, 12*(3), 257-278.

Miles R. E., & Snow C.C. (1978). *Organizational strategy, structure, and process.* NY: McGraw-Hill.

Mintzberg, H. (1979). *The structuring of organizations.* NJ: Prentice - Hall.

Morgan G. (1986). *Images of organization.* London: Stage.

Mulej, M., Potocan, V., Zenko, Z., Kajzer, S., Ursic, D., Knez-Riedl, J., ... & Ovsenik, J. (2004). How to restore Bertalanffian systems thinking. *Kybernetes: The International Journal of Systems & Cybernetics, 33*(1), 48-61.

Nohria, N., & Berkley, J. D. (1994). An action perspective: The crux of the new management. *California Management Review, 36*(4), 70-92.

Peters, T. (1987). *Thriving on chaos: Handbook for a management revolution.* NY: Knopf.

Pfeffer J., & Salancik G.R. (1978). *The external role of organizations: A resource dependence perspective.* NY: Harper & Row.

Planck, M. (1945). *Treats on thermodynamics.* NY: Dover.

Ruelle, D. (1988). Can non-linear dynamics help economists?. *Santa Fe Institute Studies in the Science of Complexity*, (5), 195-204.

Russell, B. (2009). *Principles of mathematics*. Routledge.

Sarmah, H. K., & Paul, R. (2010). Period doubling route to chaos in a two parameter invertible map with constant Jacobian. *Int J Res Rev Appl Sci*, *3*(1), 72-82.

Scott, W. R. (1981). *Organization roland, natural and open systems*. NJ: Prentice-Hall.

Simon, H. A. (1962). The architecture of complexity. *Proceedings of the American Philosophical Society*, *106*(6), 101-123.

Sirtes, D., & Oberheim, E. (2006, November). Einstein, entropy and anomalies. In *AIP Conference Proceedings* (Vol. 861, No. 1, pp. 1147-1154). American Institute of Physics.

Smircich, L. (1983). Concepts of culture and organizational analysis. *Administrative Science Quarterly*, 339-358.

Smuts, J. C. (1926). *Holism and evolution*. NY: Macmillan Company.

Spagnolo, F. (2003). Fuzzy logic, fuzzy thinking and the teaching/learning of mathematics in multicultural situations. In *Proceedings Int. Conference. on Mathematics Education in the 21st Century* (pp. 17-28).

Stacey, R. D. (1992). *The chaos frontier: creative strategic control for business*. Butterworth-Heinemann.

Starbuck, W.H. (1976). Organizations and their environments. In M. Dunnette (Ed). *Handbook of Industrial and Organizational Psychology* (pp. 1069-1124). Chicago: Rand McNally.

Terrybery, S. (1968). The evolution of organizational environments. *Administrative Science Quarterly*, *12*(2), 590-613.

Thompson J., D. (1967). *Organization in action*. New York: McGraw-Hill.

Toffler, A., & Toffler, H. (1995). *Creating a new civilization: The politics of the third wave*. Turner Pub.

von Bertalanffy, L. (1950). The theory of open systems in physics and biology. *Science*, (111), 23-9.

Yager, R. R. (1992). Entropy measures under similarity relations. *International Journal of General System*, *20*(4), 341-358.

Zadeh, L. A. & Desoer C.A. (1963). *Linear system theory*. NY: McGraw-Hill.

Zadeh, L. A. (1965). Fuzzy sets. *Information and Control*, *8*(3), 338-353.

Zadeh, L. A. (1975a). Fuzzy logic and approximate reasoning. *Synthese*, *30*(3-4), 407-428.

Zadeh, L. A. (1975b). Calculus of fuzzy restrictions. In *Fuzzy sets and their applications to cognitive and decision processes* (pp. 1-39). Academic press.

Zadeh, L. A. (1990). The birth and evolution of fuzzy logic. *International Journal of General System*, *17*(2-3), 95-105.

ABOUT AUTHOR(S)

YAVUZ ERCIL is professor in the Department of Public Relations and Publicity at Başkent University. He received his Ms degree from Istanbul University, Institute of Business Administration, Department of Management and Organization. His PhD degree was from Gazi University, Institute of Social Sciences, Department of Management and Organization. He works in the fields of strategic management, simulation, network science and system analysis. Since 2017, he is a member of the board of directors of Başkent University Center for Strategy and Technology.

MERVE KADAN completed her undergraduate education at Yıldırım Beyazıt University, Department of Engineering and Natural Sciences, Electrical and Electronics Engineering. Since 2019, she has been contuining her Ms degree at TOBB University. Her interest includes studies on Fuzzy logics and Multi Criteria Decision Making.

3

SYSTEMS THINKING BASED STRATEGY DEVELOPMENT

UFUK TUREN

> *"Whosoever desires constant success must*
> *change his conduct with the times"*
> *– Machiavelli*

ABSTRACT

Strategic management is known as an overall effort to create, shape, manage the future of an organization with all aspects and stakeholders to sustain and flourish. Short sighted and linear approaches in strategy development efforts cannot help in establishing a robust strategy to ensure organizational sustainment and competitive advantage since this type of approaches likely leads to many different individual and organizational level learning and understanding traps. Moreover, ever-increasing volume of complexity and related data inside and outside of organizations also aggravates the challenges within the organizational decision and steering processes. At this rate, sensing, collecting, storing the relevant data from all required sources in an organization and its environment, storing them in a proper way, make them ready to be soundly retrieved and used in properly designed integrated decision support tools gain importance. In this chapter, to mitigate those needs listed above;

firstly, the rationale behind the necessity of employing systems thinking in strategy development is discussed using theoretical bases and exemplification with some classical organizational strategy related archetypes. Secondly, the transition phases of systems thinking use in strategy development is shortly visited. Lastly, the criticality of strategy management systems and their evolution is scrutinized from systems thinking perspective.

INTRODUCTION

Strategic management has many different definitions causing from its nature. It can basically be defined as an effort pre-emptively steer the future of an organization with its all shareholders (higher, lower, sides) in order to maintain and progress its competitive advantage in its environment which is currently considered as a domain of infinite competition (Zahn, 1999).

Competitive strategy of an organization basically identifies the potential services/products and markets, long-term objectives, and plans and policies to achieve the stated objectives. Identifying the features of market priority, product structure, manufacturing configuration, and investment preferences are viewed as indispensable and continuous procedures for an organization, which is eager to have and sustain competitive advantage (Errin, 2004). Competitive advantage is mainly regarded as somehow balanced and positive state which is fed by organization's two complementary aspects namely **attractive positions** in its environment and/or its **distinctive resources** such as capabilities, competences, knowledge bases, etc. Organizational competitiveness relies on its ability to the successful development of new ways and strategies, which rests upon its capability to unceasingly generate new sources to sustain its business advantages in times of volatility caused by rapid change and high competition (Zahn, 1999).

Organization is a system consisting of people, forms, structures, technologies, operations and processes that function together to make and keep itself alive. Systems thinking principally looks at the whole as a basis for understanding, designing, and managing its components that have convergence with management strategy and implementation. Since the middle of 20th century systems thinking approach is used in organizational management discipline. Although its applications have been focused on operational and tactical levels in the beginning, strategic level efforts of organizations' managements imported it into different organizational and functional domains (Rajagopal, 2012: 33) to gain insight with regard to holistic consistency, coherency, and future planning requirements.

Systems thinking, deriving from systems theory, posits that a system should be distinguished from its environment by drawing a boundary to the system's internal components and processes. Inside its boundary, the system has a certain level of integrity ensuring the subsystems are working together coherently and this integrity level offers the system interior to boundary a degree of independence. The systems interconnect with their environment using inputs and outputs such as data, information, knowledge, energy, substances, commodities and services. The difference between the inputs and outputs is defined as the function committed a system and the value added by the organization. Moreover, thanks to interactions among subsystems, the system can own some abilities and characteristics that its parts do not own (Weissenberger-Eibl, Almeida, & Seus, 2019). Figure 1 shows a basic form of a system.

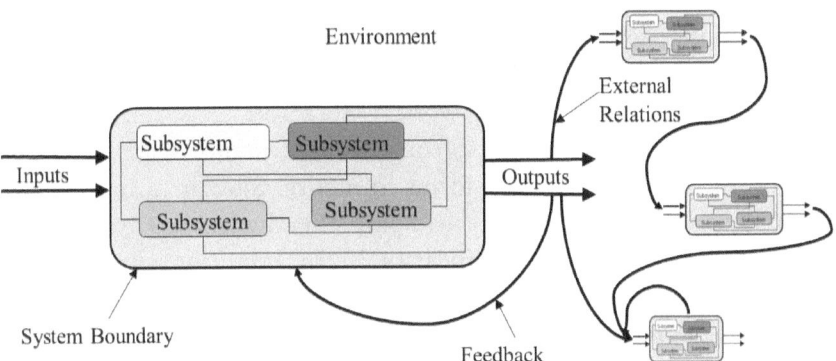

Figure 1: Basic Elements of Systems and Environment

The main aim of this chapter is to provide insight into the theory and practicality of the systems thinking approach in organizational strategy development processes using the basics of strategy management systems and fundamental methods of systems thinking with some examples. However, touching the issue of organizational learning always seems us the crux of all matters discussed in this chapter. Therefore, we begin with it.

RATIONAL ORGANIZATIONS WITH SYSTEMS THINKING

When the systems thinking perspective is implemented to the strategic management domain of an organization, it is seen that the organization having different subsystems with various functions (e.g., procurements, R&D, marketing, sales, etc.) is surrounded by a multifarious environment (i.e., competitors, local/international market, dealers, customers, governments, politics, etc.), and it can be posited that organization's consistency and coherence with those internal and external factors determines the success.

Avoiding the Linearity Trap and Understanding the System

Organizations' classical understanding of linear environmental conditions is another perception trap mostly caused by human's inclination to see things in linear ways, steady, smooth, and tidy. Humans feel more comfortable in a linear and deterministic environment with certainty and their mind often prefer risk aversion and uncertainty avoidance. This approach is also used profoundly in organizations in order to foment orderliness in workplace and ensure higher efficiency and productivity (Bratianu, 2015). By doing so, they keep themselves in a secure environment saving them from complexities of the real world (Morecroft, 2015: 32-33). The preferred and common form of manmade structures can simply be distinguished from ingenious forms in nature. Similar to individuals, organizations have a tendency to comprehend their environment in linear forms. Although nonlinearity is the common characteristic of real-life inherited from the nature (for details see Wohlleben, 2018) this compelling understanding also holds in academia and industry. Especially in management and economics field, most of the empirical studies are based on experiments, which can only provide linear associations between different variables (Bratianu, 2015). This deterministic linearity delusion causes people to assume that the linear functions between two or more different variables last forever at the same pace. When it is scrutinized in terms of cause and effect of the events or processes, this linear approach brings us a type of thinking named as "Event-oriented Thinking" and defined as a hypothesis suggesting that complications are occasional and randomly emanating from unpredictable and irrepressible happenings in the environment.

Variety of occasions is infinite. There is no need and no time to worry about the causes of a problem since they drop out of the sky suddenly. Fixing the problem itself, rather than tackling with the causes, as soon as possible is the only and most significant course of action. In this way of thinking, the association between problem and solution is perceived as linear and deterministic since it repels the causes and focuses on palliative remedies. The problem undertaken is seen as a divergence between a critical collective end and an undesirable or unfavorable current state. This can conduce to rapid, conclusive, and determined actions (Morecroft, 2015: 32-33). This open loop, firefighting fashioned interventions are mostly reactive instead of proactive. Senge (1990) denotes this problematic thinking pattern as **fixation on events**. This thinking habit inclines to believe that everything can be linearly explained by causal chains of events. The root causes are the actions triggering the chains of cause and effect from event X and Y to Q (Figure 2).

However, fundamentally there is no linear relationship between variables. It is just an oversimplification of relation in a point in time. An increase in the level of one variable can merely induce an increase or a decrease in the level of the other for a limited period. In time, the linear function becomes obsolete to explain the association and the necessity of developing nonlinear functions emerges eventually. Understanding the nonlinear nature of interactions between variables is an enlightening step not only for individuals but also for organizations. In this point the systems thinking concept emerges as a powerful tool providing individuals and organizations with ability to grasp nonlinearity and complexity by using an integrative view where the whole is more vital that its components and often bigger than the sum of them. Many simple correlations and even some complex associations between variables (e.g. inputs, outputs) can be traced. The complexity of holistic structure and behaviors makes the analysis not only more difficult but also more precise (Bratianu, 2015). In systems thinking the behavior of a system emerges from the structure of its feedback loops. Root causes are not individual nodes. Instead, they are the forces emerging from particular feedback loops. Figure 2 shows the basic difference between linear thinking and systems thinking approaches.

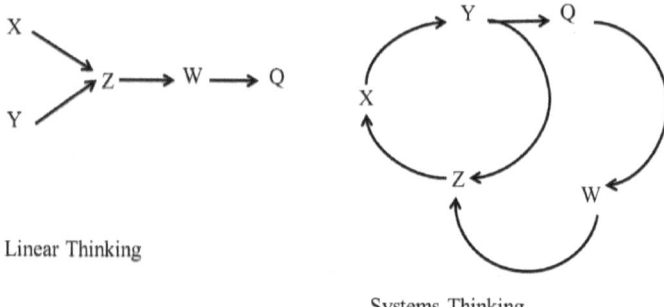

Linear Thinking

Systems Thinking

Figure 2: Linear Thinking versus Systems Thinking

Dynamic Complexity and Poor Inquiry Skills

One of the reasons exacerbating policy resistance is the complexity of the world dominating and hampering our humanly understanding. Humans' mental models are often restricted, internally inconsistent, and unreliable. Our ability to grasp the evolving impacts of decisions is incompetent. We make decisions and take pertinent steps based on our temporary, insular or biased reasoning which is also called as **bounded rationality** (Simon, 1957: 198) but due to our inadequate appreciation of complexity, eventually these decisions often coming back and disturb us. Comprehending and dealing with the causes of policy resistance necessitate understanding not only the complexity of the system but also the current mental models dominating the organization and directing decision making processes. Despite most individuals relate complexity only with the number of elements or subsystems in a system or number of options to be deliberated to reach a conclusion, dynamic complexity may emerge even in simple systems with low numerical complexity. Especially for counterintuitive behavior of social systems it may rooted from the interaction of actors over time (Sterman, 2001: 21). **Bullwhip effect** which was first formalized by Forrester (1961) can be given as an example for this phenomenon. The term was devised to delineate the response of the demand-supply mechanism to slow and small variability in consumer demand with hefty fluctuations in the dealers' inventories and production for suppliers through the supply chain. This is analogous to the knob of a bullwhip causing heavy blows at the popper (Wang & Disney, 2016).

Time-consistency/Sensitivity

Delay is a reality in cause and effect processes in real life. A time delay in a causal connection between an outcome and its pertinent cause(s) are observed (Morecroft, 2015: 20). Contrary to most people's expectation there is almost no process having immediate link between cause and effect. There is always at least a slight lag in cause and effect relations even if we cannot perceive it with our humanly receptors. For example, if an object moves slow enough to be unperceived by human eye, it can easily be deemed as stationary. This beguilingly perceived inertia might cause different consequences bad or good but definitely unexpected. Although its trueness has been debated by some scholars (Fallows, 2006), the fable describing a frog which is slowly boiled alive is a common story being told as an example to visualize the unnoticeable slow motions or change in circumstances and their unexpected implications (Senge, 1990: 22). The actions of frog affecting the temperature of water is fractional and is mostly ignored in this example. However, the existence of a feedback from frog's body to the water temperature cannot be denied. This phenomenon is a type of perception trap sometimes called as **invisibility of feedback** or **hidden feedback**.

Missing the causes and consequences of delay can bring people and organization in extreme chaotic states in their effort to control tools, devises, behaviors, and/or systems, and usually leads to fluctuation effects such as instability and oscillation in system behavior (Sterman, 2000). Delays are often seen in business. For example, investing in a new product or process today is committed to be utilized in the future. It takes time for a new worker to get up to speed and catch the desired efficiency and productivity level. There is always some delay in seeing any benefits following the resource allocation to a particular project. Although these types of delays are usually very well acknowledged and tolerable, they can be extremely hazardous if organizations fail to recognize or possibly underestimate them (MBD, 2014; Eakes, 2018).

Consistency

Components of an organization do tactical and operational goal and objectives definitions. Alas, stove pipe approach in different and each branch and divisions provides and pushes them achieve their own objectives alone (Morecroft, 2015: 37). However, some systems archetypes such as **limits to growth** and **success to the successful** or **tragedies of the commons** just decelerate their speed and gradually halt them as they approach the boundaries of other components and their unconsidered individual goals (Senge, 1990; Braun, 2002); Bellinger, 2004). This phenomenon can only be tracked and dealt using an overall big picture approach and awareness initiatives. Theoretically, it is posited that the stovepipe mentalities, as well as narrow and mindless departmental/functional standpoints characterized by seeing ourselves as separate (**the enemy is out there** syndrome) from the world are often nurtured (mostly unintentionally) by organizations (form-behavior) and by human's tendency to slice the facts and problems for analysis (Senge, 1990: 40-42; Morecroft, 2015: 37). Goal and objective definition efforts of subsystems should be in line with organization's holistic and strategic objectives. **Strategizing process** starts here.

Organizations cannot exist, be defined, or understood without their environment. Irrevocable interactions with the other agents in the environment posit this reality and this holds for all living organism besides organizations. Open systems view of general systems theory (von Bertalanffy, 1950) reveals this intermingled symbiotic, competitive and **coopetitive** way of coexistence. Imponderable nature of exogenous variables has been always a source of unexpected circumstances in environments where organisms or organizations eager to survive. Some of these unexpected conditions emerge from some phenomena, which are not sensitive to the behavior of the individual, organism, or organization, or the sensitivity link is too long and connectivity is too low to observe the implications on the environmental conditions just in time. In fact, interconnectedness nature of everything in the nature is valid for every case (Wohlleben, 2018). Delays in feedback lines just make them invisible to naked human eye. For example, an ostrich with its head in the sand is often used metaphor for individuals or organizations mostly insensitive the changing nature of their environment (Volokh, 2003).

Understanding the Basic Behaviors in a System

Stocks and the flows are known as the main construction blocks of real-life systems. They are fundamental in various field of study from zoology or anatomy to transportation or thermodynamics. For instance, a population of a mammal species increases by births and decreases by deaths; the affliction of alcohol in human body increases by intake and decreased by exudation. The journey and conversion of substance among the locations and states respectively is crucial for the dynamic nature of complex systems. In applied and pure sciences domain, resources are habitually palpable such as the level of sodium in the blood, the number of homeless people in a country (Sterman, 2006). In social sciences, employee skills, organizational commitment, and motivation can be listed as examples of intangible resources. Research denotes two distinct reasons leading people's poor innate appreciation of stocks and flows. The first is called **narrow mental boundaries** meaning that individuals are not usually mindful of to the meshes of flows and stocks causing accumulation and depletion resources. Sterman (2006) gives a congruent case of California's Air Resources Board as an example for this ignorance. California's Air Resources Board pursues to control and diminish air contamination by encouraging the use of zero emission motoring. Although electricity or hydrogen powered motoring does not have a fuming exhaust pipe, it is a fact that the electricity or hydrogen plants do contaminate the air. This understanding just displaces the emission location to somewhere remote, but in the planet, and potentially harmful for others or next generations.

The second is called **stock-flow fallacy** and denotes people's poor intuitive comprehension and discrimination between stocks and flows in the process of accumulation. Inflow and accumulation of any resource in any system is perceived and assumed as correlated by the majority of people. For example, people likely consider that the deficit in government budget increases the total national debt and envisage a positive linear association between them. However, in reality, a stock still continues to increase during the decreasing behavior of the net inflow, as long as the net inflow is not negative because of the integral relation between them. This means that a debt of a government rises while the budget deficit goes down until it reaches the zero, becomes negative and turns into a budget surplus. Then, the depth starts to decrease (Sterman, 2006).

Understanding the Policy Resistance (Defensive Routines)

Traditional organizational structures and management practices are threatened by harsh competition, diminishing growths in productivity, sudden technological, political, and environmental change, and disappearance of market and national boundaries. In this turmoil the fact that organizations must change more rapidly than before is broadly admitted. Organizations trying to survive under these pressures require striving to elucidate their mission, vision, and principle values. Voluminous seek to recalibrate into leaner, more locally controlled and market responsive structures. Alas, the core operating policies guiding organizational behavior often remain unchanged though initial expectation is otherwise.

One of the reasons core policies remain unchanged is that the thinking underlying such policies remains unchanged. The **dominant logic** syndrome (Bettis & Prahalad, 1995) and the **defensive routines** (Sterman, 2000: 32) behavior are denoted as other reasons for the policy resistance in organizations. In this situation, the crux is that once an organization has established an efficacious strategy, it spares no efforts to preserve it. Sincere efforts to infuse new management practices often incur frustration, discomfort, and cynicism rather than necessary and enduring improvements. Efforts to develop new and groundbreaking strategies often plummet because new strategies and pertinent novel organizational forms and structures ruffle prevalent traditional ways, standards, and suppositions. The problem lies here with inability to identify the criticality of dominant mental models. However, new strategies are apparently the product of new worldviews or paradigms. Insightful change in the strategy can only follow enormous change in the ways of thinking. Indeed, improving the mental models of managers is considered the fundamental task of strategic management (Senge & Sterman, 1992).

In short, unexpected side effects can be generated by emerging policies that attempt to provide betterment for the system from the policy owner's perspective. The efforts aiming to stabilize an organization can destabilize it. The decisions made and policies applied may be perceived as harmful by the other agents and incite negative reactions. Those others may want to keep the previous state of the system or restore the balance we disrupted. This reactive state in social systems is also called **counterintuitive behavior of social systems** often results in policy resistance, the inclination, and the impact of interference to be hindered, attenuated, or overthrown by the reaction of the system or a set of subsystems to the interference effort itself (Sterman, 2000: 11). **Policy resistance** is a type of balancing feedback in systems thinking. Figure 3 shows a simple example of systems thinking model of policy resistance. Arrows indicate causation; actions aiming immediate correction can alter the initial mindset or environmental conditions. The signs (+) and (-) denote positive and negative causations respectively. Repeatedly implemented immediate fix method often has a long-term side effect. This side effect loop changes the initial assumptions or rules and in time the opted immediate fix loses its effectiveness and becomes futile. Policies promising short time bright solutions often accumulates long term disappointments or indignations. After a certain delay, accumulated side effects change the game and policy owner sees that the once upon a time working method is obsolete or ineffective.

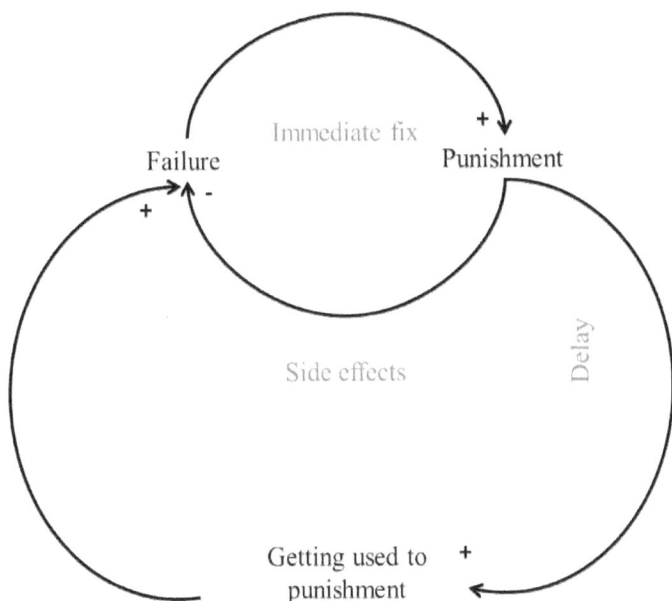

Figure 3: Policy Resistance (Fixes that fail)

BEING A LEARNING ORGANIZATION TO DEVELOP SOUND STRATEGIES

Believing in the understanding that scientific experimentation had to be invented is not easy since humankind learns naturally by trial and error by trade, and it seems a sensible way to explore the world and universe by testing hypotheses against observation. Governing our thoughts and tethering them to what the world teaches us, seems the best way to get on with discriminating what is true and what is false (Garvey & Stangroom, 2012: 213). For the historical and anthropological evolution of curiosity, suspicion, exploration, experimentation and empirical inclination of human please see e.g. Ferguson (2011) and Harari (2014).

Learning is the ability to change, which is stimulated by sensing, observing, information processing, deciding, and stepping forward or take action coherently. All living organisms have somewhat learning capability in their pertinent lives and environment. Organizational learning is considered as a significant intangible ability to change and survive in ever-changing and even volatile environment. In an increasingly complex business environment, an organization having no ability to learn cannot survive for long. Either it loses most of its labor since the other players in the industry or broader environment creating better conditions for employees and establish much more attraction while the non-learner keeps its initial obsolete settings or it produces an excessively large amount of products without noticing the shrink in demand or keeps taking orders from dealers or customers without checking supply-in chain or production department. Organizational learning not only consists of these first-order learning cycles, second and third-order learning cycles can be practiced by highly mindful organizations (Sterman, 2000: 18). As it seen in the Figure 4, single-loop (first order) learning loop is about perceiving the facts and react in proper ways whereas double-loop (second order) learning loop denotes reconsidering the ways of thinking such as paradigms and developing new decision rules and strategies.

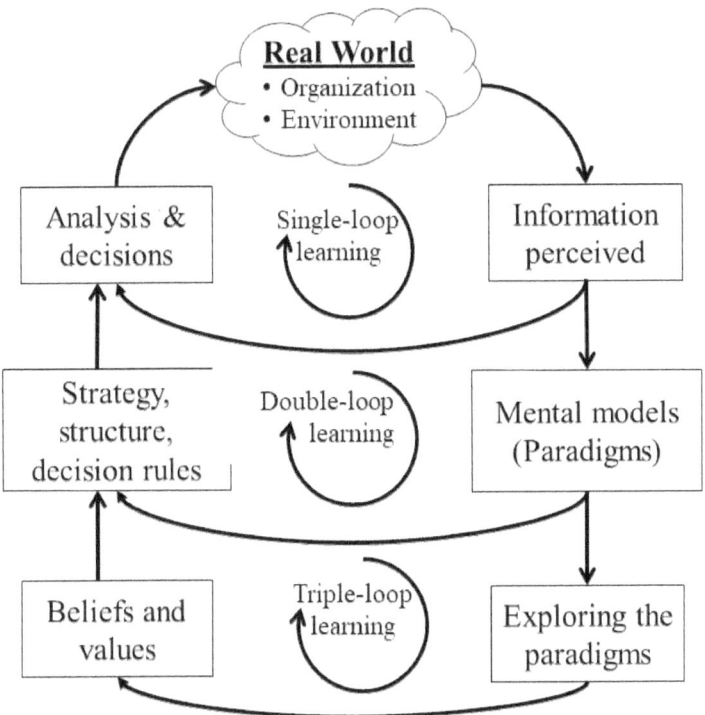

Figure 4: Learning is a Feedback Process
Source: Adapted from Sterman (2006)

Single-loop Learning

The concept of feedback is the basic generator of all dynamics and learning behavior in a system. As individuals perceive divergence between the desired and real circumstances, they do something that, as they believe, will induce the real world to converge to the coveted circumstance. Latest observation or intelligence gathered on the real-world circumstance leads them to review their insights and engender them to make decisions for the next step (Sterman, 2006). The apparatus of thermostat operating in one mode is a commonplace example plied to elucidate this type of learning. Once it senses that the room temperature is below the defined threshold, it turns on the boiler. Otherwise, once it notices that the temperature is above the threshold, it turns off the boiler. The rules are kept without questioning.

Double-loop Learning

In double-loop learning, individuals or organizations can question the rules on whether they should be amended if the circumstances necessitate and contemplate on how to design the corrections in rules. More **thinking outside the box**, critical thinking and creativity naturally are more commonplace in this type of learning. This learning often supports individuals or groups to understand the reason behind the supremacy of a certain solution over others in addressing an issue or reaching a target. Double-loop learning is considered momentous for the organizational wellness and performance, particularly in times of hasty changes or volatility. Double-loop learning takes place when people in an organization engage in discussion about their reasons for failure on quality expectations, and whether the expectations were realistic or not. Amendments in quality expectations or recalibration of quality inspection systems might be the outcomes of the discussion session. Individuals have got the opportunity of understanding much more about themselves, others and organization using their perceptions, beliefs and values. As an addendum, **triple-loop learning** describes the learning about double-loop learning. It is practiced when, after having engaged in a session in which a discussion occurs about the reasons and possible solutions of some issues, the group discuss the dynamics, form and quality of the session in terms of the way it is conducted and the extend of usefulness for learning process (Crites et al., 2009). The triple-loop learning has not been reflected in Sterman's (2006) model since it is likely considered a phase of double-loop learning.

Organizational Learning Based on Motivational Feedbacks

Learning dynamics of organizations should be apprehended to grasp their evolution and environment as well as their competitive advantage and relative strengths (Dickson, 1996). Dicksons, Farris, and Verbeke (2001) posit few general types of learning dynamics that can generate constructive feedback interactions within and between organizations, namely; motivation, learning by doing, localized learning, learning ability, surveillance, and routinization dynamics. To keep it short and sharp without skipping the critical aspects, two-group categorization is preferred to be mentioned here namely "within-organization and between-organizations learning" dynamics. It should be accepted that the competing motivations between and within organizations habitually show short and long-term effects.

Within the organization, motivation dynamics can usually occur through **try-harder** feedbacks, which is known as a type of reinforcing loops (R). For example, deterioration in employee benefits and wages triggered by an exterior factor such as a financial crisis or a belt-tightening policy after a handover of organization's ownership or a general election. The shrink in employee welfare benefits likely results in a drop in employee morale and motivation. Decline in morale and motivation is highly epidemic and presumably causes a fall in employee service quality and/or quantity. This often reduces the demand for the organization services or products. Because of the drop in revenues due to the fall in demand, further belt-tightening measures apply to reduce the costs causing another drop in employee benefits. This mechanism goes on spiraling down as a vicious cycle. The nature of this vicious cycle is very similar to the reinforcing cycle, depicting a positive pulse in employee benefits that increases morale and enthusiasm, service performance, demand for organizations products and services, and then benefits for employees (Dickson, Farris, & Verbeke, 2001). Exogenous factors can be very influential on this reinforcing (R) positive feedback loops (Figure 5).

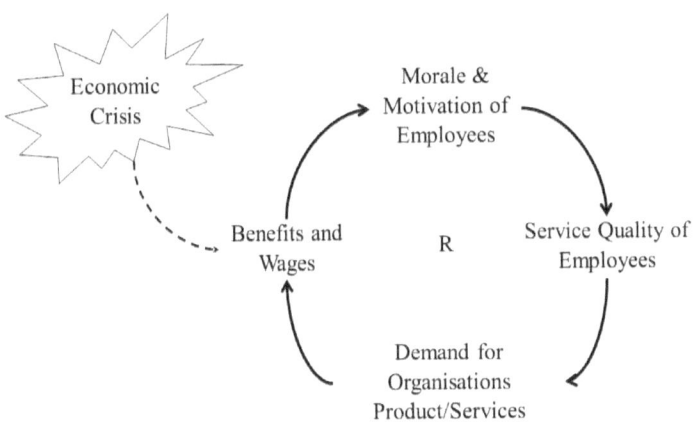

Figure 5: Reinforcing Loop Example Fragile to an Exogenous Shock

In learning-by-doing cycle, the more organizations acquire the more they function and execute tasks. Then, the better they function the better they learn. Moreover, customers learn from using or consuming the products and services provided. The more they consume the more they get used to it and learn. The more they learn the better they utilize and reach a higher usage utility level. This is also called **experience learning** functions in all processes and industries. The more employees do and learn the more efficiency and productivity occur. This provides positive impact to overall utilization and economies of scale in organizations. Consumers' learning cycle could be very important determiner when they are exploited in a systematic approach. For example, Singer initiative of home sewing machines marketing effort was a bright idea and far beyond of its era. It was the first machinery producer to resort to advertising demonstration to let the public know about its products by organizing a distribution network of travelling training teams to teach women how to sew and use its their state-of-art machines (Nersesian, 2000: 44). This effort initiated the learning process at consumer's side and within a decade, mothers were teaching daughters, and sisters were teaching sisters how to sew using Singer machines. This endeavor resulted in an autonomous and reinforcing customer learning-network, which aided Singer to dominate the global market and to withdraw from training-focused distribution network strategy and engage more in mess and standardized sales and distribution strategy without sewing machine training since the women's familiarity and experience were functioning flawless. The company's success in global markets is mostly attributed to its innovations in marketing (Godley, 2006).

The Microsoft is another classic example sharing almost the same dynamics. As the producer of an innovative computer operating system and office software providing unprecedented non-technical user involvement into computers world, Microsoft conquered the market easily and has dominated it up today. The more popularity the software gains in business world more successful the software becomes because the workers who are familiar with the software can help to train new ones to use the software (Dickens, Farris, & Verbeke, 2001). Following the accumulation of enough capable users in the environment, the organization does not need to invest more on innovative marketing efforts. After that moment, the mechanism is habitually similar to **path dependency** theory. Path dependency describes the sustained use of a product, procedure, or service rooted in historical preference or use. An individual or an organization may keep on use of a product, process, or service albeit newer and better substitutes are available in the market. Path dependency happens since it is often easier or cheaper to stay along an already established path than establishing a novel one (Banton, 2019). Today, majority of the households, schools, and organizations in the planet function freely and decisively on behalf of Microsoft as the consumer training-network as well.

The archetype **price wars** is one form of aggressive competition has been observed between organizations in the real world likely leads to aggressive reactions. This mechanism can cause either an ever-increasing revengeful competition or an advantageous cooperation. Although it can cause total destruction for the competing organizations, over time it also serves as a beneficial cycle for innovation efforts struggling to find better ways to increase relative quality while decreasing prices to survive as well as customers enjoying the higher quality with cheaper prices. In the automobile industry, the quality improvement motivation connected to the competition motivation cycle can also reinforce itself as competitors make every effort to beat each other in fewest reported defects by users or market share, and provide quality accumulation and spillover effects in all shareholders in the environment in the long term. This mechanism can lead to a phenomenon called **The Red Queen Effect** positing that the preference of customers enjoying the competitive cycle between organizations change over time. This is a result of price and quality wars over the longer run. Competing organizations learn from the innovators and re-innovate to supply other new segments in the environment or find emerging profitable areas to exploit being coherent with the changes in customer preferences fed by two-way learning between organizations submitting products and services, and their customers. This reinforcing cycle has been called the Red Queen Effect since the Red Queen of Alice and Wonderland told Alice that she had to run ever faster just to keep the same position against rivals as they were striving to outpace. In this context, the competitive environment is considered a favorable ecology of learning organizations (Barnett, 2008: 50-52). Although, this growth of rivalrous motivation to learn and improve performance is credited as an essential aspect of free market economies, running faster using the dominant strategy mostly does not provide enough merit to stay competitive (Voelpel, Leibold, Tekie, & Von Krogh, 2005). Figure 6 displays two Systems Thinking models to explain the Red Queen

Effect in a competitive environment. The reinforcing loop model at right is the product of two balancing loops (B) model, as each organization (A & B) is keen to achieve desired objective and slightly surpass the opponent in each iteration or cycle. As the advantage is passed back and forth between the two parties, a reinforcing loop (R) emerges due to this escalating interaction of two balancing loops.

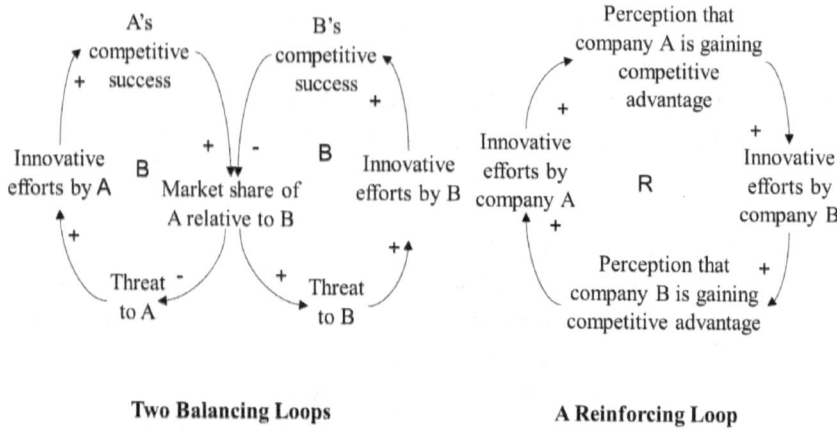

Two Balancing Loops **A Reinforcing Loop**

Figure 6: The Red Queen Effect by Systems Thinking

In majority of the free markets, the culture of innovation has been established by competing actors and their pertinent supply chain tails. Not only the primary actors heavily challenging each other but also the others large and small supporting the operations are this nonstop race of sustainability and even survival. In the first sight these two critical sentences might sound repulsive. However, the competing factors are the basic reality of the nature including the basic building blocks of living organisms: genes (Dawkins, 1989; Gardner & Welch, 2011). Thus, development, betterment, progress, or improvement processes are tightly dependent on the competition not only in the domain of pure science but also in that of social science. In this sense, organizations also learn from each other while they compete. This cooperative competition paradox is called as **coopetition** (Bengtsson, Raza-Ullah, & Vanyushyn, 2016). Inter organizational learning-networks boost the likelihood of both learning by observing and direct training. The breadth and depth of network's absorptive capacity emerges based on the openness, honesty, collaboration, and coordination inclinations or climate in a network. Studies show that learning networks among rival organization are generally useful to increase awareness, stimulate inspiration, and boost even imitation, and drives its members to localized learning trap (e.g. Kotabe & Swan, 1995; Bower & Christensen, 1995) that often hinders exploration and favors exploitation or localized learning.

The strategy development in organizations has conventionally been an occasional effort, which is either triggered by crises or remembered in periodic cycles. It has been a process highly monopolistic and concentrating more on the analysis procedure rather than an exercise of fusion since it has been often handled by a few subject matter experts inside the organization alone and/or a group of outside consultants. As a result, managers at different levels have not been able to develop strategic thinking competence. This deficiency would probably result in failure in turbulent rather than stable conditions. In this context, the effectiveness of a chosen strategy will not only be determined by the content of it, but also by flexible adaptation ability gained during the strategy development process. This ability is shaped in concordance with the anticipated actions and reactions of other actors in ever-changing environment during the strategy development exercise. A dynamic perspective fed by this mechanism is necessary for organizations to disentangle from sticking with once upon useful but now obsolete strategies. Strategy development process should be based on various factors such as situations inside and outside the organization, tangible and intangible actualities, etc. (Zahn, 1999).

RESOURCE BASED APPROACH TO DEVELOP STRATEGIES USING SYSTEMS THINKING

The main raison d'être for all organizations is to survive and keep their competitiveness. To that end, they endeavor to improve quality, reduce cycle times, increase yields, and maximize throughput and lower costs for their business processes by employing available resources. A restrictive or singular attention on improving the cycle time, throughput, quality or cost generally does not entail distinct competencies (Norton & Kaplan, 1999: 93). Most organizations understand the importance of building and conserving the necessary resources of their business, both tangible items such as capital, raw product, investment, employees and customers and intangibles such as employee morale and motivation, investor support or customer loyalty. It is also clear that resources are mutually dependent since a low delivery service damaging organizational reputation vitiates the value of good product quality. Similarly, the poor quality of a product or service cannot be saved by a well-motivated salesforce. Thus, ranking the resources in order is almost futile because the entire organization can be imperiled if any one of them is in poor state. Likewise, determining value drivers in an organization is often a recipe for failure since the best after sale service would be unrewarding for angry customers who paid money for a range of products that repetitively flop or services often does not comply with deadlines (Warren, 2002: 15).

An organization's resources contribution to competitive advantage can be found in many Strategic Management texts under the name of **resource-based view** of strategy. It can be observed that the fundamental significance of organizational resources has been increasingly in the research agenda for more than 60 years (Penrose, 1959; McDougall, Wagner, & MacBryde, 2019). It looks fairly obvious that resources are imperative to sustain organizational performance over a period, rather than just construing it at present. In most of the research, the nature of this delayed dependency, in terms of how to quantify the impact of each separate resource on organizational performance measures such as profitability or market share or the interdependencies between various resources, is not apparent (Warren, 2002: 17). In the classical resource-based view of strategic management, the main focus is on the durability of resources, mobility or tradability, replicability, substitutability and complementarity of resources. Figure 7 displays the classic linear understanding of strategic resources and their relations to reach a performance measure (Return on Investment) for an organization, which is quite linear and in a cascade form without the feedback effects and time concept. On the other hand, Figure 8 depicts the situation once the time dimension and feedback interrelations are added.

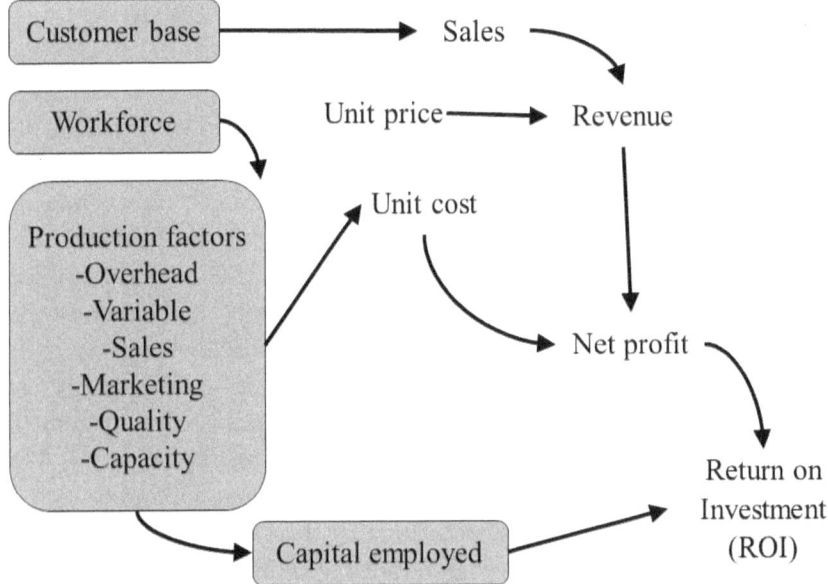

Figure 7: Linear Relations of Strategic Resources and Organizational Performance

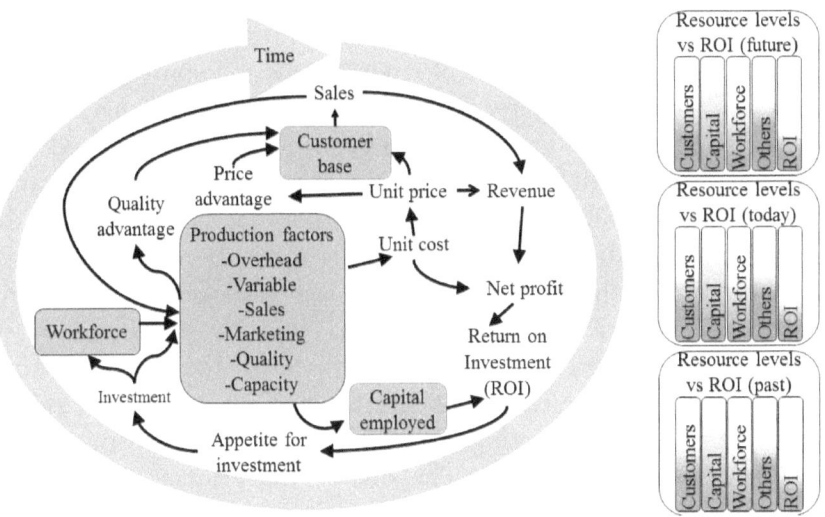

Figure 8: Nonlinear relations of Strategic Resources and Organizational Performance with Time Dimension

STRATEGY DEVELOPMENT AND STRATEGIC PLANNING

Strategy is claimed to be originated from the Greek word **stratego** meaning to plan the annihilation of one's adversary through the effective use of available resources. The notion was industrialized purely in relation to the efficacious and realistic chase of victory in warfare and stayed as a military term until the19[th] century once it inaugurated to be used in the civilian business world, nonetheless, most writers consider the genuine course by which this took place is not an easy one to be traced (Bracker, 1980; Chandler, 1962; Clegg, Schweitzer, Whittle, & Pitelis, 2017: 6). When Drucker (1954) first set forth strategy and strategy development as a methodology to organizational management, his effort was to define the business domain of an organization. This concept attracted little attention until 1962 when Chandler (1962) recognized the significance of coordinating the various traits under one all-embracing strategy in his revolutionary work **Structure follows Strategy** defining strategy and framing the process by which it could be operationalized. Based on Drucker and Chandler's work, Ansoff (1965) and Andrews (1971) strongly contributed to this growing strand of field. They explored the real business needs and scrutinized the rapid alterations of environmental circumstances.

As matter of fact organizations, insensitive to their pertinent industry, function in a turbulent environment and consequently have to be calibrated properly for long-term survival. Within this concept, strategy is recognized as the linkages between organizations and their larger environment. Hence, the choice of the most appropriate strategy for a given organization is very vital. The decisions defining the strategies are predisposed by numerous factors such as organizational form/structure, the characteristics of top management team and board, organizational culture/climate and available resources. The opted strategy subsequently influences the organizational performance. The strategic management domain principally focuses to comprehend the contingent implications of opted strategy on organizational performance. The one of the primary concerns or the most critical research topics in contingency theory literature is to identify the variables related to the preferred strategy and the form of organizations and to examine their impact on organizational performance (Doty, Glick, & Huber, 1993; Kariuki, Awino, & Ogutu, 2011). Organizations are always dependent to and influence their environments meaning there are two ways interactions between them (Drucker, 1954; Chandler, 1962; Ansoff, 1987). Through strategic planning, organization can investigate and acquire from their environments, establish strategic goals, generate strategies in order to reach or at least move in an agreed direction and satiate numerous stakeholders (Mintzberg, 1973; Porter, 1987). In general, strategic planning consists of three main components: strategy formulation, strategy implementation, and strategy evaluation (Ansoff, 1987; Bailey & Johnson, 2001; Mintzberg, 2008).

Campbell and Alexander (1997) assert that strategy has nothing to do with plans or planning tasks, but it relates to insights. Alas, strategy is generally used with planning as a phrase namely strategic planning. Strategy development is not strategic planning (Zahn, 1999). Planning is a procedure of analysis and programming aiming to establish time-based milestones and sequenced pathway towards the visions and/or long-term objectives already exists and dictated by strategy development process or strategizing (Mintzberg 1994; Zahn, 1999; Johnson, Scholes, & Whittington, 2008: 399). Insight, intuition, and sagacity, based on solid supporting evidence, reinforce strategy development processes as the principle products of strategizing. In an increasingly intricate and volatile business environment, organizations are required to reassess their strategic decisions on a regular basis. As global ambiguity and puzzling complexity raises, the question of how to manage these challenges becomes more prevalent for organization. Strategic planning attempts to offer companies the necessary means and methods to prepare for and defend successfully against unceasing alterations in their environment (Grant, 2003; Kiliko, Atandi & Zachary, 2012).

Strategic management literature is rife with a myriad of methods to be practiced by organization in order to develop their strategies (e.g. Bryson, 1988; Hesterly & Barney, 2010; Hitt, Ireland, & Hoskisson, 2012; Hubbard, Rice, & Galvin, 2014). Most of those recommended strategy-developing methods share a few basic phases and steps. The conceptualization phase in which the strategic objectives are described and verbalized. The development phase in which multiple options or course of actions are formulated and assessed, and the preferred course of action is then carefully chosen as a strategy. The implementation phase comprises of the execution and monitoring the selected strategy (Grainger-Brown & Malekpour, 2019).

Systems Thinking, the Centre of Strategy Developing

Systems thinking inspires us to discover inter-relationships, perspectives and boundaries. Systems thinking is principally valuable in addressing complex or extraordinary problem situations. These problems mostly cannot be resolved by any one actor, and a complex system cannot be fully understood from only one perspective (Allen, 2019). The systems thinking approach in essence provides stimulating pathway to identify them using some stereo type learning traps or archetypes (Colleen, 2018). Mostly dealt with analytic and linear ways, classical strategic management and strategy developing approaches have been reformed and enriched using systems thinking ways and tools for a few decades Thus, systems thinking is often used interchangeable with strategic thinking or the core element of the strategic thinking (e.g. Lyneis, 1999; Andersen, Bryson, Richardson, Ackermann, Eden, & Finn, 2006; Haines, 2009).

Phases of Systems Thinking in Strategy Developing

Increasingly complex business environment drives organizations to habitually reconsider their strategic choices. However, organizations are besieged by inability to competently gather and construe with pertinent information and knowledge on their environment. Even if they collect and analyze information with regard to their competitors, value chain shareholders or customers, they often ignore the influence of broader environment such as society or governmental formations and institutions since they are not able to evaluate their strategy development procedures through systemic perspective and posit an attitude of picking information bits discriminatorily. Organizations need to acquire the ability to see themselves as a part of and closely interrelated with the mesh of environmental complexity. The systematic screening of environment should be practiced concurrently with a detailed and careful investigation of organization's interior conditions such as competencies and/or positions and advises a holistic framework as a measure to develop an insight to apprehend organizations as systems. The scenario technique backing systematic exploration of organization and its environment offers useful recommendations for managers in order to support them to incorporate a phased systems thinking methodology into strategy development journey. This approach to strategy development practice has four outcomes namely (1) an effective understanding, structuring, and analysis of organizational problem or decision point, (2) education of the management in the dynamics of their business to keep them conversant with the process and ensure their active to participation in the structuring of the problem and in the analysis and interpretation of findings, (3) selling of others within the organization who are not involved in the project, on the recommended course of action, and (4) providing the means for permanent learning and planning with the model for longstanding exploitation and success (Lyneis, 1999). To reach these outcomes, Lyneis (1999) proposes a four-phase methodology, which is a modified and refined version of Morecroft (1985) and Richmond (1997) and claims his approach strikes the balance between **product** and **process**. Figure 9 shows the phases and their pertinent cost (time, money) and benefits (accuracy, validity).

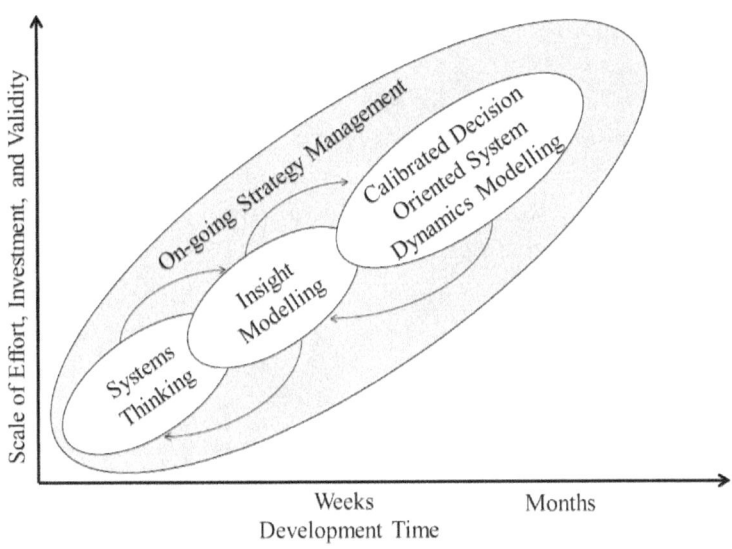

Figure 9: Phases of Systems Thinking in Strategy Development

Phase 1: Business Structure Analysis

The first phase clearly describes the problem of concern, the possible sources of that problem, and binding restraints affecting the implementation of any likely resolution. The performance drivers of the organization and possible solutions are aimed to be detected. In this phase, organizational documents, procedures, and available data are reviewed. The opinions of managers, pertinent customers and competitors are taken into consideration. The questions such as "What induces consumers to purchase this product? What induces them to purchase from one supplier rather than another? What encourages the internal acquisition and the decision of resource allocation?" are focused. The major environmental factors likely affecting the business are revealed. This phase is heavily relied on the tools and techniques known as systems thinking and delivers behavior-over-time graphs, causal-loop and stock-flow diagrams using related system archetypes and mental simulation.

Phase 2: Developing a Small, Insight-based Model

The system dynamics modeling, as a powerful tool of systems thinking consists of building and investigating computer-based mathematical models. This is a two-step phase. The first step denotes the construction of a relatively small intuition-based model to describe the basic dynamics of the processes under examination and engendering more understandings to determine the necessary course of actions to enhance the performance of organization. The relatively tiny model is useful for the education and training of the organization about the system dynamics modeling. In the second step, the first model is improved to reach a relatively more detailed version and calibrated using historical data. The quantification of advance model aims to absorb all relevant knowledge and experience available to increase validity, persuasiveness, and plausibility.

Phase 3: Developing a Detailed, Calibrated Model

During the final phase the construction and calibration of a more detailed model is the main focus. This phase aims to guarantee that the model covers the required variables and associations to form and validly represent the system behavior under investigation. Conceptual models and even small, insight-based models can be incomplete or misrepresent features or forms which are considered dynamically important. The accuracy of model in this stage is critical to the soundness and reliability of the results, which will likely be input to strategic decisions. Organizations often invest in this expensive effort to facilitate their strategy development abilities. Thus, the findings of the analyses such as the experimentation of different course of actions, exploring sensitivities of various stakeholders, etc. based on the model are likely used in strategic level decisions. The ownership of the journey is another critical aspect for the success. The key decision makers should involve the process from the beginning not only as a contributor but also as a champion.

Phase 4: On-going Strategy Management with Systems Thinking

The entire strategy management process is tackled in three interrelated domains, namely analysis, planning, and control. **Analysis** is typically activated by a noteworthy and/or obstinate deviance between actual and anticipated level of outcomes. It necessitates configuring, examining, and enhancement of an organization's appreciation of its strategic complications and evaluation of the possible and feasible course of actions available to deal with the undesired situation in the behavior of the system causing performance gap. During the process of evaluating alternative course of action, as the basis for new strategies, the strategic level problems was faced or likely will be faced by the organizations often becomes apparent and triggers further investigations.

Planning comprises of the assessing the feasible options, selecting and implementing the best strategies for an organization. Evaluation of competing options is subject to the projected goals of the organizational and the actual level of performance. The standing strategies and goals require refinement, when necessary, coherent with changing conditions inside and outside organization. Traditional, episodic strategy management approach often focuses only on analysis and planning. Once the strategic course of action is opted and implementation is triggered, the strategy development process turns off up until another unexpected and undesired substantial trouble occurs. Strategic level **control** phase is usually overlooked and waiting until a substantial gap accumulates at the performance indicators. Control is a critical factor for a permanent systematic performance monitoring and rapid and accurate feeding back of achievements, failures, issues, potential risks and threats. A robust control function for strategy management necessitates continuous and organized performance tracing and the efficient feeding back of achievements, opportunities, complications, perils, risks, prospects, experience, knowledge, and lessons learned to the other modules of the strategy management effort.

CHALLENGES IN USING SYSTEMS THINKING IN STRATEGY DEVELOPMENT

Objective, controlled, and replicable experiments are considered necessary in order to grasp reliable and valid evidence to discriminate among competing hypotheses or course of actions. However, the ability conduct and practice these comprehensive efforts become more difficult as the phenomenon go more complex. Real life strategies and policies inlays in complex networks of physical, biological, ecological, technical, economic, social, political, and other associations. Experimentation efforts in intricate social systems are usually considered immoral, infeasible, extremely expensive, or difficult to practice replications. Thus, decisions made in/for one part or at/for a particular period. Decisions made in one sub-systemic domain demise across the systems because of insensitivity and mindlessness exacerbated by the topographical, disciplinary, and time-based boundaries across the systems. Because of long-time delays the full consequences of today's actions will never be experienced (Sterman, 2006). It is necessary to understand the major hurdles standing in the way of the systems thinking use in strategy developing process. The first challenge is access to data and applicable methods to analyze data. The **integrated longitudinal databases** are required to retrieve data and analyze them using proper techniques to test dynamic hypotheses. Today, in spite of the developments in big data technologies and availability of such databases, a narrow selection of numerical and mostly financial records is often kept and provided by only some of the organization. The data capturing the hard and soft variables of organizations for long enough time is necessary for the precision of the analyses and reliability of the conclusions. Robust methods to analyze those longitudinal data such as **panel data analysis** has gained more importance over cross-sectional studies. The findings of those longitudinal data analysis methods are often more appropriate for systems thinking models. The second challenge is the impact of **multiple organizational dimensions** (e.g., individuals, groups, and organizational subunits) and supraorganizational factors (e.g. customers, competitors, governmental actors, and social norms) on the organization's dynamic behavior under examination. To reach more realistic and valid conclusions, multi-firm, multi-national, multi-cultural models may be required to develop sound strategies through employing various scenarios and analyzing the sensitivities under the bubble of near real-life conditions. The third challenge can be tackling multiple-level problem sets. For decades, academia has kept theories simple enough to be analyzed and assessed in slices. However, today the explanatory power and predictive validity of them are broadly questioned. Nearly decomposable world vision does not help anymore. Complexity of interactions and interrelatedness of various different variables from numerous disciplines necessitates an ability to integrate those diverse domains to provide scientific precision and validity. The fourth challenge is to understand and reflect human behavior and decision-making process in an organization. Those soft variables are considered critical for the performance of organizations and strategy-developing efforts fall short without them (Jackson, 2003: 101). For example, innovative methods and technologies ensuring to transfer tangible variables to intangible or vice versa in a definitive function evolving in time are required to capture longitudinal data on soft variables. Moreover, capturing and importing data with different structure and translate them to be adaptable and coherent in the same model as tangible and scalable variables could be very welcomed.

ENTERPRISE STRATEGY MANAGEMENT SYSTEMS

In an ever more intricate business environment, organizations regularly necessitate reviewing their strategic picks and associated assumptions. Under the pressure of growing global uncertainty and perplexing complexity, managing these challenges has become a significant issue for most organizations. Organizations employ strategic planning to be effectively prepared for constant and somehow volatile environmental changes (Kiliko, Atandi, & Zachary, 2012; Grant, 2003). Because of this volatile, unpredictable and instable character of globalized business environment both problems and their solutions have gradually become ephemeral. Therefore, the necessity of decision support systems having ability to learn and reconcile swiftly has increased dramatically (Weissenberger-Eibl, Almeida, & Seus, 2019). Making strategic decisions during strategy development always requires information and expertise from not only each different department and function of an organization but also the other partnering actors across the supply chain, competitors in the environment, customers, and the governmental situations. As an accurate and robust database with different analytic capabilities, **enterprise information systems** are used to provide required on-time information support for operational and strategic level decisions and are considered a high priority for the top level managements of organizations (Turban, Aronson, Liang, & McCarthy, 2007: 411). For decades, the need for these integrative information management systems has been fulfilled by materials requirement planning (MRP) yesterday and enterprise resource planning (ERP) systems today.

As the continuation of MRP systems, ERP systems are an information system integrating organizational functions and processes to add a holistic value and decrease costs by providing the right information to the right subsystems or individuals at the right time to support them in making better analyses and decisions in allocating organizational resources efficiently and pre-emptively. They consist of multi-module software applications interrelatedly functioning and supporting multiple varieties of functional areas of an organization. These stupendous networked, computerized, all-business, cross-functional systems are configured to engender higher organizational efficiency and effectiveness. ERP systems typically supports the functions such as accounting, manufacturing, human resource management, purchasing, inventory management, inbound and outbound logistics, marketing, finance, and engineering (McGaughey & Gunasekaran, 2007). Today ERP systems continue to evolve with the efforts of highly competitive ERP companies towards to an Internet-based architecture coherent with the ever-increasing prevalence of e-commerce and the globalization. Most ERP systems have integrated modules such as advanced planning and scheduling, sales force automation, customer relationship management, supply chain management, and e-commerce modules/capabilities. The holistic system provides necessary data, information, knowledge, intelligence and expertise based on its embedded data acquisition, storage, organization, retrieving and surveillance abilities as well as sophisticated analytical tools to understand and track necessary performance indicators and measure. This ability provides authorities with the continuous and real-time track of each critical performance indicators.

Today ERP vendors provide a pack of tools and methods such as the **Balanced Scorecard** and **Activity-Based Management** to support the strategy management process of their customer organizations. ERP systems require both financial and non-financial measures to assess overall performance of an organization. The balanced scorecard provides a comprehensive measurement framework having four dimensions: financial, internal processes, customer, and innovation and learning (Kaplan & Norton 1992). Activity-Based Management offers the profitability picture of an organization, which is the output of the Activity-Based Costing analysis to help managers focus on improving activities having the primary effect on the net profit (Cooper & Kaplan, 1991). Although ERP vendors assert that they offer innovative best practice by providing a pack of tools and methods within ERP, their implementation is likely to be problematic in the practice of strategy development and evaluation. For example, there are various problems with the lack of a time dimension, the lack of clarity regarding the interrelationship between the different perspectives in a score case and the lack of evidence relating to causality of the measures to improve performance in Balanced Scorecard (e.g. Norreklit, 2000; Brignall, 2002; Madsen & Stenheim, 2015), and the top-down and reductionist management approach of Activity-Based Management (Johnson, 1992). More notably they have a linear and event-based understanding of the business processes and the evolution of organizational variables over time, and the interactions between actors and their behaviors inside and outside of the organization provided by feedbacks are mostly neglected. Activity-Based Management's applicability to all type of organizations in all settings is also questioned and criticized (Brignall, Fitzgerald, Johnston, & Silvestro, 1991). The Balanced Scorecard is also criticized for its failure to contemplate the impact of competitors' performance, the perspectives of employee, supplier, or other actors in the environment (e.g. Ballantine & Brignall, 1995; Kennerley & Neely, 2000). However, as the most concerning part, the problematic simplistic and linear tailored interactions analytics among the provided tools and methods are assumed flawless probably because of research orientation in business and organizational management until recently, which has been dominated by single discipline, single function, event-based, cross-sectional lines of effort (Brignall & Ballantine, 2004). Interdisciplinary or holistic research based on Systems Thinking has been very infrequent. Thus, the expectation of the organizations is coherently limited. However, lately a few pioneer ERP providers claim some integrated solutions covering soft aspects of the organizational strategic resources such as intellectual and social capital as well as hard ones (Shi & Wang, 2018).

Since organizational performance measures have enormous impact on strategic management, they should initially be shared with not only the individuals in the position to develop and implement strategies and pertinent plans but also reflected in vision and mission statements of top management to make them know for entire organization. The information system well-tailored to reflect these performance measures and providing real-time accurate data to feed the calculations and analyses can provide the strategic or operational level managers with proper decision support. However, choosing the right performance indicators among the critical resources and behaviors is challenging task especially in classic strategic management strands. Objectively measuring and control of the developed strategies is another problem of classical strategy management systems. The measures opted to assess a strategic performance depends on the organization and the objectives decided to be achieved. The objectives that were established earlier in the strategy development phase of the strategic management process (e.g. productivity, market share, better intellectual capital, and/or lower costs, etc.) should definitely be reflected in the process of measuring organizational performance following the instigation of the pertinent strategy (Hunger & Wheelen, 2003). However, performance of a strategy is the result of the entire effort and resource-based approach is considered a good option to be used in systems thinking oriented strategic management efforts by employing interrelated resource accumulating functions (Warren, 2002: 312). This approach is useful to unravel the stovepipe understanding of different functions and departments, and even the entire organization in mental trap of seeing the world linear and discrete and ignoring the effect of others inside and outside of the organization. Systems thinking and system dynamic modeling tools easily functions on the various ERP systems using integration platforms portals such as SAP NetWeaver or Oracle Cloud Platform. Although Dynaplan shows its integration certificate to SAP in its website, most ERP systems keep as priority to integrate and support different business applications to share a common database (McGaughey & Gunasekaran, 2007; Shen, Chen, & Wang, 2016).

On the other hand, a few different approaches aiming to measure the performance of strategic plans using systems thinking and systems dynamics modeling using the information flow in the network of shareholders (Baskici & Ercil, 2019) among many other cross-sectional studies attempt to explore the precedents of successful strategies (e.g. Esfahani, Mosadeghrad, & Akbarisari, 2018).

Conclusion

Strategic thinking is considered the core effort in and the most crucial input for strategy development process or strategizing. Thus, according to Christensen (1997: 156) it should be cultivated as a core competence and kept inside the organization, instead of outsourcing internally or externally, with genuine ownership, full engagement, and support of top management. Top management should take personal responsibility voluntarily and actively in the process of developing the key strategies, which will provide guidance to all in the organization. Managements should be aware that changes in strategic level necessitates new understanding, paradigm shift maybe, pushed by evolving parameters or variables inside or outside, by capturing not only hard data or information from all sources but also grasping soft insights from experiences such as tacit knowledge or wisdom. Systems thinking which is used interchangeable with strategic thinking in the literature (e.g. Zahn, 1999; Dicksons, Farris, & Verbeke, 2001; Bonn, 2001; Batra, Kaushik, & Kalia, 2010) offers necessary thinking and reasoning skills for strategic thinking. Strategies, which are expected to be better, are developed by strategy forum conducting strategic thinking skills mostly lacking systems thinking approach. As a result, strategies to promote new practices often fail or worsen the current problems they are anticipated to resolve.

Organizational learning based on systems thinking likely prevents such policy resistance, but slow and weak learning is commonplace in the case of complex social systems. Complexity often hampers organizations' ability to realize the delayed and remote aftermaths of taken actions, and inadvertent **side effects**. Moreover, the process of learning usually founders even in the face of strong evidence. Typical mental models generate flawed but self-verifying suppositions, which are conducive for bringing about detrimental beliefs, attitudes, and behaviors. This steadfast vicious mechanism, driven by ideology, superstition, or unconscious bias without precise testing, mostly undermines the implementation of useful and creative strategies to flourish.

Lacking of seeing and understanding the big picture of actors, resources, behaviors and interactions, organizations usually are trapped in the first order learning loop hampering creativity and initiative, stunting the development of the skills with regards to second and third order learning. Using delays, stocks and flows, feedbacks systems thinking offers a collection of tools assisting organizations to grasp deeper and better awareness and understanding on the forms of systematic interactions and problems from humble pen-and-paper methods such as causal loop diagrams to more intricate tools such as system dynamics simulations. System archetypes provide a bunch of templates very useful to capture the **common dynamic vignettes**, which frequently occur in varied circumstances and configurations. They are influential tools for identifying complications and spotting critical and influential leverage points, which may trigger and support key changes and change management efforts in an organization.

Organizations should recognize the significance of a fundamental shift in the way they generate, be aware of the delayed, distal, and nonlinear effects of their strategies using systems thinking approach in their strategy management efforts. Seeing the big picture is king. The reductionist inclination of specialization, departmental segmentation, or social stratification will be no longer useful to resolve contemporary problems. Although it often promotes higher enthusiasm, deeper understanding and finer subject matter expertise in a specific functional domain in an organization, alas, it feeds up the defensive routines by narrowing the boundaries of individual's or organizationally shared mental models. Likewise, categorization of organizational problems based on functions may pose ownership and non-ownership stratification risk in organizations. Indeed, all integrated and interdependent by nature, they are in the boundary of the organization (Meadows, Randers, & Meadows, 2004; Sterman, 2006). Thus, establishing boundaries are not useful for solving the problems affecting and affected by all. Nonetheless, some boundaries are needed and indispensable to simplify the complexity and focus on objectives and hypotheses in systems thinking methodology, the management of an organization need to be aware that drawing nonessential boundaries often cause organizations not to recognize some critical feedbacks and foster overconfidence of decision makers about their ability to control environment and others sharing the environment. This state leads organizations to create others once they solve one problem.

In developing strategies, generally, it is impossible, infeasible, or too expensive to conduct experiments in the real world to yield objective and robust evidence to support decisions. In those cases, virtual worlds and simulation environments such as system dynamics come forward as a relatively reliable method to test dynamic hypotheses about the strategies developed, evaluate their prospective impacts, and explore how intricate dynamic mechanisms work and evolve.

Systems thinking based strategy management and system dynamics modeling integrated with organizations main enterprise database frame can also provide real-time evidence to support strategic decision-making processes by increasing accuracy and reducing the decision cycle times. This integrated approach offers an environment conducive for developing more robust, timely and holistic strategic management and measuring systems fed by systems thinking and/or systems dynamics modeling and simulation. Accuracy and shorter decision cycles are critical organizational abilities for a competing organization in any industry. As you remember from Red Queen Effect model, the side having agile and faster moving ability directs the reinforcing loop on its own behalf.

REFERENCES

Allen, W. (2019). *Systems thinking*. Retrieved from https://learningforsustainability.net/systems-thinking/ on 09.09.2019.

Andersen, D. F., Bryson, J. M., Richardson, G. P., Ackermann, F., Eden, C., & Finn, C. B. (2006). Integrating modes of systems thinking into strategic planning education and practice: The thinking persons' institute approach. *Journal of Public Affairs Education, 12*(3), 265-293.

Andrews, K. R. (1971). *The concept of corporate strategy*. Homewood, IL: Dow Jones-Irwin.

Ansoff, H. I. (1965). *Corporate strategy*. New York: McGraw-Hill.

Ansoff, H. I. (1987*). Corporate strategy*. 2nd Ed. London: Penguin Business.

Bailey, A., & Johnson, G. (2001). A framework for the managerial understanding of strategy development. In Volberda, H. W., & Elfring, T., (Eds.), *Rethinking strategy* (pp. 212-230). London.

Ballantine, J., & Brignall, T. J. (1995). A taxonomic framework for performance measurement. In *Proceedings of the 18th Annual Congress of the European Accounting Association*. Birmingham, UK.

Banton, C. (2019). *Path dependency*. Retrieved from https://www.investopedia.com/terms/p/path-dependency.asp on 01.12.2019.

Barnett, W. P. (2008). *The red queen among organizations: How competitiveness evolves*. Princeton University Press.

Baskici, C., & Ercil, Y. (2019). In pursuit of information: Evaluating strategic plans. *VINE Journal of Information and Knowledge Management Systems*. Retrieved from https://www.emerald.com/insight/content/doi/10.1108/VJIKMS-03-2019-0037/full/html on 25.12.2019.

Batra, A., Kaushik, P., & Kalia, L. (2010). System thinking: Strategic planning. *SCMS Journal of Indian Management,* 7(4).

Bengtsson, M., Raza-Ullah, T., & Vanyushyn, V. (2016). The coopetition paradox and tension: The moderating role of coopetition capability. *Industrial Marketing Management, 53*, 19-30.

Bonn, I. (2001). Developing strategic thinking as a core competency. *Management Decision, 39*(1), 63-71.

Bracker, J. (1980). The historical development of the strategic management concept. *Academy of Management Review, 5*(2), 219-224.

Brătianu, C. (2015). Developing strategic thinking in business education. *Management Dynamics in the Knowledge Economy, 3*(3), 409-429.

Braun, W. (2002). The system archetypes. Retrieved from https://www.albany.edu/faculty/gpr/PAD724/724WebArticles/sys_archetypes.pdf on 15.12.2019.

Brignall, S., & Ballantine, J. (2004). Strategic enterprise management systems- New directions for research. *Management Accounting Research, 15*(2), 225-240.

Brignall, S., Fitzgerald, L., Johnston, R., & Silvestro, R. (1991). Product costing in service organizations. *Management Accounting Research, 2*(4), 227-248.

Brignall, T. J. (2002). The unbalanced scorecard: A social and environmental critique. In A. Neely, A. Walters, R. Austin (Eds.), *Performance measurement and management: Research and action* (pp. 85-92). Cranfield, UK: Centre for Business Performance, Cranfield School of Management.

Bryson, J. M. (1988). A strategic planning process for public and non-profit organizations. *Long Range Planning, 21*, 73-81.

Campbell, A., & Alexander, M. (1997). What's wrong with strategy. *Harvard Business Review,* Nov.-Dec., 42-51.

Chandler Jr., A. D. (1962). *Strategy and structure: Chapters in the history of the American industrial enterprise.* Cambridge: MIT Press.

Clegg, S. R., Schweitzer, J., Whittle, A., & Pitelis, C. (2017). *Strategy: Theory and practice.* 2nd Ed. London, UK: Sage.

Colleen, L. (2018). *The vocabulary of systems thinking: A pocket guide.* Retrieved from https://thesystemsthinker.com/the-vocabulary-of-systems-thinking-a-pocket-guide/ on 10.10.2019.

Cooper, R., & Kaplan, R. S. (1991). Profit priorities from activity-based costing. *Harvard Business Review, 69*(3), 130-135.

Crites, G. E., McNamara, M. C., Akl, E. A., Richardson, W. S., Umscheid, C. A., & Nishikawa, J. (2009). Evidence in the learning organization. *Health Research Policy and Systems, 7*(1), 4.

Dawkins, R. (1989). *The selfish gene.* Oxford University Press.

Dickson, P. R. (1996). The static and dynamic mechanics of competition: A comment on Hunt and Morgan's comparative advantage theory. *Journal of Marketing, 60*(4), 102-106.

Dickson, P. R., Farris, P. W., & Verbeke, W. J. (2001). Dynamic strategic thinking. *Journal of the Academy of Marketing Science, 29*(3), 216-237.

Doty, D. H., Glick, W. H., & Huber, G. P. (1993). Fit, equifinality, and organizational effectiveness: A test of two configurational theories. *Academy of Management, 36*(6), 1196-1250.

Drucker, P. (1954). *The practice of management.* New York: Harper and Row.

Eakes, S. (2018). *Managing delays. The system thinker.* Retrieved from https://thesystemsthinker.com/managing-delays/ on 01.09.2019.

Errin, E. (2004). Technological intelligence and competitive strategies: An application study with fuzzy logic. *Journal of Intelligent Manufacturing, 15*(4), 417-429.

Esfahani, P., Mosadeghrad, A. M., & Akbarisari, A. (2018). The success of strategic planning in health care organizations of Iran. *International Journal of Health Care Quality Assurance, 31*(6), 563-574.

Fallows, J. (2006). *The boiled-frog myth: stop the lying now!. The Atlantic.* Retrieved from https://www.theatlantic.com/technology/archive/2006/09/the-boiled-frog-myth-stop-the-lying-now/7446/ on 27.09.2019.

Ferguson, N. (2012). *Civilization: The West and the rest.* Penguin.

Gardner, A., & Welch, J. J. (2011). A formal theory of the selfish gene. *Journal of Evolutionary Biology, 24*(8), 1801-1813.

Garvey, J., & Stangroom, J. (2012). *The Story of philosophy: A history of Western thought.* Quercus Publishing.

Godley, A. (2006). Selling the sewing machine around the world: Singer's international marketing strategies, 1850–1920. *Enterprise & Society, 7*(2), 266-314.

Grainger-Brown, J., & Malekpour, S. (2019). Implementing the sustainable development goals: A review of strategic tools and frameworks available to organisations. *Sustainability, 11*(5), 1381.

Grant, R.M. (2003). Strategic planning in a turbulent environment: Evidence from the oil majors. *Strat. Mgmt. J., 24*, 491-517

Haines, S. (2009). *Strategic and systems thinking. The winning formula.* Haines Centre Book Series on Business Excellence, San Diego: Haines Centre for Strategic Management.

Harari, Y. N. (2014). *Sapiens: A brief history of humankind.* Random House.

Hesterly, W., & Barney, J. (2010). *Strategic management and competitive advantage.* Upper Saddle River, NJ: Pearson/Prentice Hall.

Hitt, M. A., Ireland, R. D., & Hoskisson, R. E. (2012). *Strategic management cases: Competitiveness and globalization.* Cengage Learning.

Hubbard, G., Rice, J., & Galvin, P. (2014) *Strategic management: Thinking, analysis, action.* London, UK: Pearson.

Hunger, J. D., & Wheelen, T. L. (2003). *Essentials of strategic management (Vol. 9).* NJ: Prentice Hall.

Jackson, M.C. (2003). *Systems thinking: Creative holism for managers.* Chichester: Wiley.

Johnson, G., Scholes, K., & Whittington, R. (2008). *Exploring corporate strategy: Text & cases*. London, UK: Pearson Education.

Johnson, H. T. (1992). It's time to stop overselling activity-based concepts. *Strategic Finance, 74*(3), 26.

Kaplan, R. S., & Norton, D. P. (1992). The balanced scorecard: Measures that drive performance. *Harvard Business Review, 70*(1), 71-79.

Kariuki, P. M., Awino, Z. B., & Ogutu, M. (2011). Effect of firm level factors, firm strategy, business environment on firm performance. *Business Environment Journal*, 1-27.

Kennerley, M., & Neely, A. (2000). Performance measurement frameworks-a review. In A. Neely (Ed.), *Performance measurement—Past, present and future*. Cranfield University.

Kiliko J, Atandi B., & Awino Z.B. (2012). Strategic planning in turbulent environment: A conceptual view. *DBA Africa Management Review, 3*(1), 73-89

Lyneis, J.M. (1999). System dynamics for business strategy: A phased approach. *System Dynamics Review: The Journal of the System Dynamics Society, 15*(1), 37-70.

Madsen, D.Ø., & Stenheim, T. (2015). The balanced scorecard: A review of five research areas. *American Journal of Management, 15*(2), 24-41.

MBD (Madden Business Development) (2014). *Delays-impact of delays in systems thinking*. Retrieved from https://www.maddenbusinessdevelopment.com/blog/building-blocks-of-systems-thinking-impact-of-delays/ on 12.10.2019.

McDougall, N., Wagner, B., & MacBryde, J. (2019). An empirical explanation of the natural resource-based view of the firm. *Production Planning & Control, 30*(16), 1366-1382.

McGaughey, R. E., & Gunasekaran, A. (2007). Enterprise resource planning (ERP)- Past, present and future. *International Journal of Enterprise Information Systems, 3*(3), 23-35.

Meadows, D., Randers, J., & Meadows, D., (2004). *The limits to growth: The 30-year update.* Chelsea Green.

Mintzberg, H. (1973). *The nature of managerial work.* New York: Harper and Row.

Mintzberg, H. (1994). The fall and rise of strategic planning. *Harvard Business Review. Jan.-Feb.,* 107-114.

Morecroft, J. D. (2015). *Strategic modeling and business dynamics: A feedback systems approach.* 2nd Ed. UK: John Wiley & Sons.

Nersesian, R. L. (2000). *Trends and tools for operations management: An updated guide for executives and managers.* Greenwood Publishing Group.

Norreklit, H. (2000). The balance on the balanced scorecard—A critical analysis of some of its assumptions. *Management Accounting Research, 11*(1), 65-88.

Norton, D.P., & Kaplan, R. (1999). *The balanced scorecard: Translating strategy into action.* Boston, Massachusetts: Harvard Business School Press.

Penrose, E. (1959). A resource-based view of the firm. *Strategic Management Journal, 5*(2), 171-180.

Porter, M. E. (1987). *The competitive advantage of nations: Strategic analysis.* Free Press.

Rajagopal, D. (2012). *Systems thinking and process dynamics for marketing systems: Technologies and applications for decision management.* USA: IGI Global.

Senge, P. M. (1990). *The fifth discipline. The art & practice of learning organization.* New York: Doupleday Currence.

Senge, P. M., & Sterman, J. D. (1992). Systems thinking and organizational learning: Acting locally and thinking globally in the organization of the future. *European Journal of Operational Research, 59*(1), 137-150.

Shen, Y. C., Chen, P. S., & Wang, C.H. (2016). A study of enterprise resource planning (ERP) system performance measurement using the quantitative balanced scorecard approach. *Computers in Industry, 75,* 127-139.

Shi, Z., & Wang, G. (2018). Integration of big-data ERP and business analytics (BA). *The Journal of High Technology Management Research, 29*(2), 141-150.

Simon, H. A. (1972). Theories of bounded rationality. In C. B. McGuire & R. Radner (Eds.), *Decision and organization* (pp. 161-176). Amsterdam: North Holland Publishing Company.

Sterman, J. D. (2000). *Business dynamics: Systems thinking and modeling for a complex world.* Boston: Irwin McGraw-Hill.

Sterman, J. D. (2001). System dynamics modeling: Tools for learning in a complex world. *California Management Review, 43*(4), 8-25.

Sterman, J. D. (2006). Learning from evidence in a complex world. *American Journal of Public Health, 96*(3), 505-514.

Turban, E., Aronson, J. E., Liang, T. P., & McCarthy, R. V. (2007). *Decision support systems and intelligent systems.* 7th Ed. Prentice-Hall of India.

Voelpel, S., Leibold, M., Tekie, E., & Von Krogh, G. (2005). Escaping the red queen effect in competitive strategy: Sense-testing business models. *European Management Journal, 23*(1), 37-49.

Volokh, E. (2003). The mechanisms of the slippery slope. *Harvard Law Review, 116*(4), 1026-1137.

von Bertalanffy, L. (1950). The theory of open systems in physics and biology. *Science, 111*(2872), 23-29.

Wang, X., & Disney, S. M. (2016). The bullwhip effect: Progress, trends and directions. *European Journal of Operational Research, 250*(3), 691-701.

Warren, K. (2002). *Competitive strategy dynamics*. England: Wiley.

Weissenberger-Eibl, M.A., Almeida, A., & Seus, F. (2019). A systems thinking approach to corporate strategy development. *Systems,* 7(1), 16.

Wohlleben, P. (2018). *The secret network of nature: The delicate balance of all living things*. Translated from German to English by Jane Billinghurs. UK: Penguin Random House.

Zahn, E. (1999). Strategizing needs systems thinking. In *Proceedings of the 17ᵗʰ International Conference of the System Dynamics Society and Australian New Zealand Systems Conference* (pp. 20-23). Retrieved from http://citeseerx.ist.psu.edu/viewdoc/download?doi=10.1.1.630.1425& rep=rep1&type=pdf on 09.09.2019.

About Author(s)

UFUK TUREN is an officer in Turkish Army, affiliated to Turkish Military Academy. He completed his undergraduate education at Turkish Military Academy, Department of Systems Engineering in 1995. He received his MS degree in Systems Engineering from Yeditepe University in 2000 and his PhD degree in Engineering Management from Marmara University in 2008. He taught Management Information Systems, System Dynamics, Decision Analyses, Decision Support Systems and Ergonomics at undergraduate and graduate levels in Turkish Military Academy, Defence Sciences Institute and Atılım University. His research interests are "human computer interaction", "technology and organizational behaviour", "macroeconomic development dynamics" and "systems thinking".

4

Systems Thinking, Cybernetics And Viable Systems Model

MEHMET HILMI OZDEMIR

> *"It is our action that determines*
> *the viability of our dream."*
> *– Steve Maraboli*

ABSTRACT

Systems thinking focuses on understanding the behavior of any system that is defined as complex structures interacting dynamically. Various classifications have been created for systems thinking that became a discipline as a result of Wiener and Bertalanffy's work at the end of the 1940s. The prominent feature of systems thinking is that it takes the system as a whole and focuses on the interaction among parts contrary to the traditional thinking. Cybernetics science that takes part under the systems thinking classification and deals with the control and management of an organization is an approach that is used in understanding the complex structures (systems). Cybernetics attempts to understand systems using the features of viability, requisite variety, and recursiveness. Likewise, feedback concept plays an important role in the science of cybernetics in the use of the information flow. Viable System Model (VSM), developed by Stafford Beer, can be defined as an approach to modeling an organizational structure based on the human nervous system. There are operational units, Meta system (management), and environment in any system from VSM perception. VSM, as a powerful functional analysis tool, has found many application areas.

Systems Thinking

System can be defined as a collection of parts that interact dynamically (Rios, 2010). **Systems thinking** focuses on simply revealing parts of complex structures and their relationships (Veeke, 2003), examining different perspectives on these complex structures, and addressing power relations and potential conflicts of interest among these aspects (Reynolds & Holwell, 2010). Senge (2011) defines systems thinking as a discipline of seeing the whole and a sensitivity that can explain the structures underlying the complex situations.

When we look at the systems thinking typology, the following different classifications are listed in the literature:

- The most commonly used trilogy classification that deals with systems thinking in a chronological order (Veeke, 2003),
- Solid Systems (operations research, system analysis, software development, database design, system engineering, etc.),
- Soft Systems (used to solving ill-defined problems),
- Critical Systems (located among sociology, organizational theory, management science and systems thinking),
- Simple and complex classification according to context,
- Classification based on system thinkers,
- The classification of systems thinking that evolved by affecting each other (Reynolds & Holwell, 2010; Veeke, 2003):
  - General System Theory,
  - System Approaches,
  - Cybernetics,
  - Information Theory.
- Classification based on chronological system approach (Rios, 2012):
  - Functionalist,
  - Interpreter,
  - Emancipating,
  - Post Modern.

Systems thinking has emerged as an interdisciplinary approach that attempts to explain systems, especially towards the end of the 20th century (Veeke, 2003). According to Reynolds and Holwell (2010); in the last 25-30 years, system approaches used in solving complex problems, which are in the foreground in terms of prevalence and impact, can be listed as follows:

 - System Dynamics - SD (Jay Forrester),
 - Viable Systems Model - VSM (Stafford Beer),
 - Strategic Options Development and Analysis - SODA (Colin Eden),
 - Soft Systems Methodology - SSM (Peter Checkland),
 - Critical Systems Heuristics - CHS (Werner Ulrich).

However, it would not be right to say that these classifications made a precise positioning on the point of view of each systems thinking. For example; VSM approach is evaluated by a group in the class of solid systems, others state that the same approach should be considered within the context of soft systems, while other researchers suggest that it is in the scope of interpretive or releasing (Rios, 2012; Reynolds & Holwell, 2010).

DIFFERENCES BETWEEN SYSTEMS THINKING AND ANALYTICAL APPROACH

The main difference between systems thinking that became a discipline as a result of Wiener and Bertalanffy's work at the end of the 1940s (Rios, 2012: 2) and traditional thinking can be explained as; **traditional thinking** is dominated by reductionist and dogmatic approaches, while **systems thinking** focuses on and prioritize an eco-system composed of relations and environment is prioritized and creative solutions (Reynolds & Holwell, 2010).

Traditional thinking takes the systems as simple structures with a static vision, and the concepts of rigidity, force, closed system, linearity and force-oriented equilibrium dominate this idea; on the other hand, systems thinking examines systems as complex structures with a dynamic vision, focusing on open system, cyclic cause-effect relationship and flow-oriented equilibrium concepts (Rosnay, 1979: 74).

While traditional thinking techniques are analytical, systems thinking techniques can be considered as **synthetic** (the phenomenon of how the components of the system work together) (Bartlett, 2001). It will not be right to try to reduce one system approach to another since both approaches complement each other with different features (Rosnay, 1979: 77). For example; the cybernetic approach within the scope of system thought points to a framework in which analysis and synthesis of systems are performed together (Hyötniemi, 2005).

Rosnay (1979: 76) states that, in the analytical approach; it is foreseen that an inference can be made with the change to be made in a variable so that the whole system can be understood, but this situation can be valid for homogeneous systems. The most important weakness of this approach is that the relations between the parts remain in the background (Bartlett, 2001) and that the system cannot be handled in a holistic way. Emphasizing that analytical approach will be weak in understanding complex systems, Rosnay (1979: 75) summarizes the differences between analytic and system approach as shown in Table 1.

Table 1: Comparison of Analytical Approach and System Approach

Analytical Approach	System Approach
Isolates the system and disassembles it and focuses on it.	Takes the system as a whole and focuses on the interaction among parts.
Examines the nature of interactions among parts.	Investigates the effects of interactions among parts on the system.
Focuses on the accuracy of the details of system components.	Cares about the holistic view of the system.
Predicts the change in a variable at a given moment.	Simultaneously, predicts the change over the group of variables.
Uses time and events in a reversible way.	Uses time and events realistically irreversibly.

Analytical Approach	System Approach
Tries to verify the facts experimentally in the theoretical framework.	Tries to verify the facts by comparing the created model with the reality.
Uses detailed and rigid models that are difficult to implement in real life.	Uses general and soft models that can be used easily.
It is effective when interactions among parts are linear and weak.	It is effective when interactions among parts are dynamic and strong.
Directs to individual discipline-oriented education.	Directs to multidisciplinary education.
Foresees application of detailed plans / programs.	Foresees goal-driven applications.
With the knowledge of the details, there are targets that are not fully defined.	Fuzzy details are available with the knowledge of goals.

Source: Rosnay (1979: 74)

In the analytical approach, the system is subdivided and focused on differences; in the systems thinking, the similarities of the sub-parts are focused on and the patterns or models produced by these parts are investigated (Bartlett, 2001). Thus, it is possible to transfer the time-dependent changes of the situations in the real world to the created models and real-time heuristic estimations can be made by adapting these models to the real world (Stewart, 2000: 101). The cybernetic approach, which is concerned with the control of real machines (electronic, mechanical, biological, economic), tries to understand the real world through the models created based on the relationships among system components (Ashby, 1957: 1).

Beer discusses analytical approach as traditional or orthodox thinking in his **Decision and Control** (1994) book, and criticizes analytical approach by comparing the cybernetic approach (Vore, 1990).

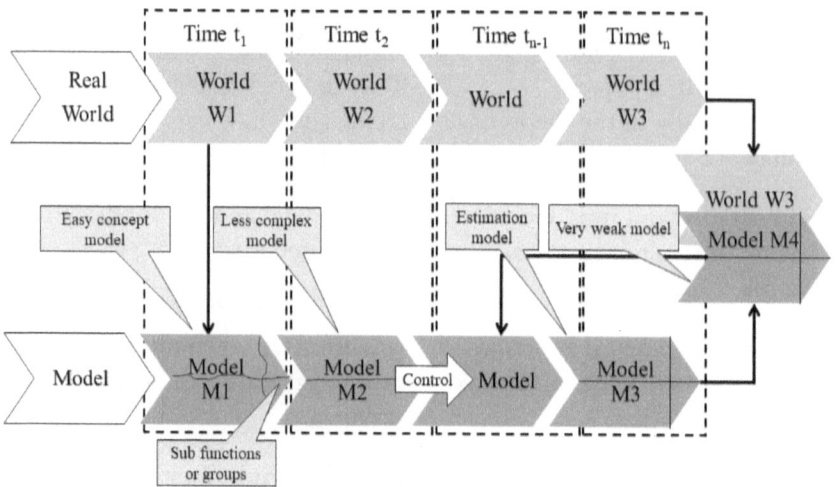

Figure 1: Traditional Control Mechanism

In the traditional approach as represented in Figure 1; first of all, the ways of reducing real world complexity are sought and an easy concept model is created. This concept model is shown as the M1 model of the real world W1 at time t_1. The M1 model is divided into sub-functions or groups (work stations, production lines, etc.) for a specified purpose. Then the complexity in the world of W1 is eliminated by computer programs to obtain the M2 model at time t_2, which is much less complex. In this case, the M2 model cannot reflect real-life attitudes, behaviors, thoughts, exceptions and relationships due to the small complexity. In addition, the M2 model tries to reflect the real world W1 at time t_1, even if it is limited, while the real world is now in the time t_2 and W2. In order to make predictions, historical data is collected to support and validate the M2 model and obtain the M3 estimation model, which is less complex than the M2 model. In the last estimation; The W4 real world is tried to be modeled with M4, which is created with a slight increase in the complexity of the M3 model. The M4 model, which is based on the M1 model at t_1 time is a very weak model for making realistic predictions that are far from reflecting the real world. In this approach, managers are left alone with a weak model away from the real world and with unnecessary data that does not produce output (Vore, 1990).

The control mechanism developed by Stafford Beer via cybernetic approach is represented in Figure 2.

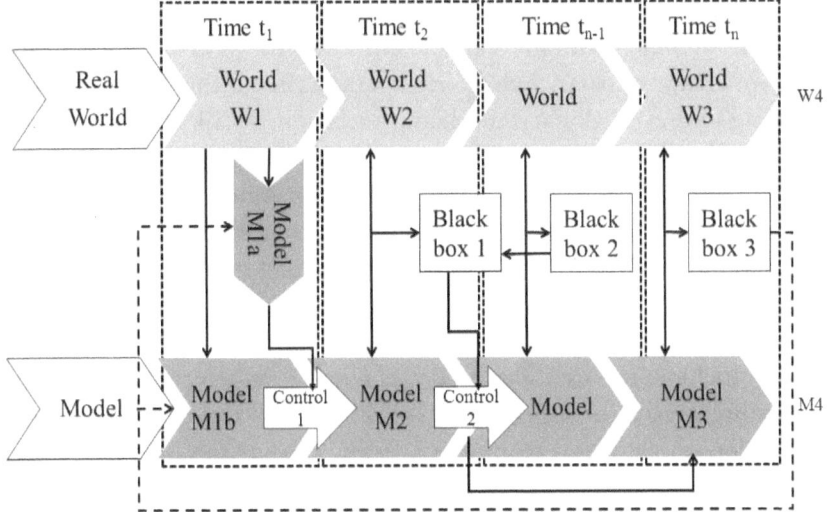

Figure 2: Cybernetic Control Mechanism

Cybernetic approach provides a management information system that produces real-time information (Vore, 1990). In the cybernetic approach, the real W1 world is reflected by the M1a structural and M1b parametric sub-models. Real world relationships are explained in mathematics, statistics and logic expressions in M1a structural sub-model. M1a can be thought of as a simple flow diagram or logic network. The sub-model M1b can be considered as a tool that serves to digitize the structural relationships in M1a. Here, it is only intended to digitize the basic relationships. In the next stage, the M2 model is created by the controller for the real world W2 at time t_2. A complex equilibrium cycle is then created between W2 and M2, and a Black Box (BB1) is placed in this loop to observe the interaction between W2 and M2. BB1 makes a statistical comparison with the M2 model by taking samples from real world situations. The controller creates the M3 model from the output of the M2 model and the output of BB1 and places BB2 in the complexity loop as in the previous step. At the same time, learning takes place by providing feedback to BB1. As a final step, the outputs of BB1 and BB2 and the output of the M3 model are utilized to generate the M4 model that predicts the real world W4 in real time. BB3 is placed in the complexity loop between W4 and M4 and feedback is provided from this black box to the sub-models M1a and M1b. The final model emerges as a self-correcting model that can reflect the real world in real time (Vore, 1990).

In the understanding and solution of problems, it is argued that linear modeling technique can be sufficient for systems with simple relations among parts, while it is emphasized that only systems thinking and evolutionary modeling techniques will be sufficient for much more complex systems (Stewart, 2000: 100).

The main reason why many problems cannot be solved is that the problem is formulated incorrectly (Beer, 1985: xiii). **Complex Adaptive Systems** have begun to replace systems that had been categorised as "simple" in the past. New approaches for solving problems in complex systems and increasing the efficiency of these systems include; the use of cybernetics science, which can be evaluated in the interpretive or soft systems thinking class, and the use of VSM, which derives its origin from this science, appears to come to the fore (Rios, 2012: 12).

CYBERNETICS

Cybernetics[4], was first used in 1834 by French mathematician and physicist Andre Marie Ampere to mean "management (la cybernetique)" (Whittaker, 2009). Norbert Wiener; during his work on automatic target detection and fire control systems, he noticed the similarity between machines and human behavior (Duffy, 1984) and discussed about cybernetics as a science in his **Cybernetics: Control and Communication in the Animal and the Machine** that was published in 1948.

Wiener defined cybernetics as communication and control science for machinery and animals (Whittaker, 2009). As shown in Figure 3; in 1948, the integration of results from different sciences and their applicability to different subjects led to a new synthetic science known as Wiener's cybernetics (Novikov, 2016).

---

4   Etymologically, it was translated into the word gubernator by the Romans, which was based on the Greek word kybernetes, which means rudder on the back of the raft, and has been reflected in the form of governor or government to today.

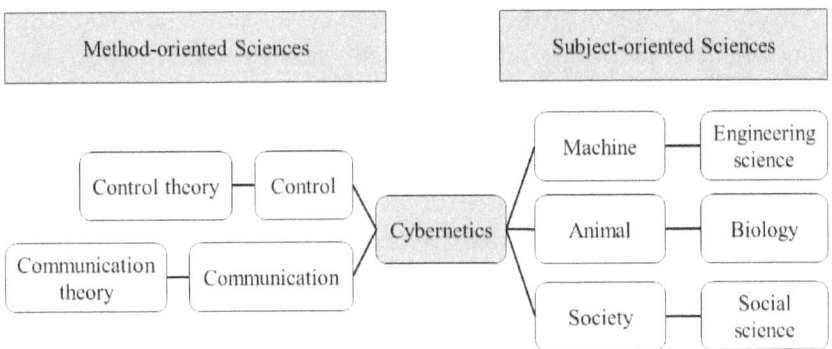

Figure 3: The Evolution of Wiener's Cybernetics
Source: Novikov (2016:10)

Beer (1985: ix) defines cybernetics as effective organizational science. Another definition took place in the literature as a branch of science that deals with the control and management of an organization (Rios, 2012: 13). Cybernetics can also be defined as communication, information manipulation and a science that deals with the control of the behavior of biological, physical, chemical and social systems (Ashby, 1957: 6; Duffy, 1984). Ashby (1957: 1) states that any behavior of systems can be discussed in cybernetics.

Although its terminology is more abstract than mathematical language, cybernetics is considered to be the most appropriate approach to solve complex problems (Schwaninger, 2006). Espejo (2013) in a research he conducted on the 2008 economic crisis; states that economists perceive the limits of computer-based mathematical models and the need for more emphasis on the paradigm of complexity and soft and behavioral factors that are evaluated beyond real equilibrium and rationality.

Cybernetics scientists created the theory of cybernetics inspired by the science of neuropsychology and used this theory in organizational modeling by comparing the features of self-management, feedback and homeostasis / ultrastability (Whittaker, 2009) in the human body (Beer, 1985: ix).

With the application of cybernetics to social systems; significant progress has been made, particularly in the concepts of information, communication, feedback and purpose (Duffy, 1984). With cybernetics, neuropsychological mapping can be made, logic and math–based models can be created to understand what each phenomenon really is (Beer, 1968).

According to Ross Ashby, cybernetics creates two important added values in the field of social systems, the first of which constitutes a common terminology for diverse systems, and the other provides a scientific method for understanding and analyzing complex systems (Duffy, 1984).

The basis of cybernetics is the knowledge and knowledge management that kills complexity according to the analogy of Beer (Juarez, Padilla, Pina, & Matamoros, 2010: 7). Beer (1985: ix), while interpreting Wiener's definition of cybernetics; stated that information is very important in management and control systems, and emphasized the recognition of feedback process is considered as a scientific invention by cybernetic scientists.

The concept of **feedback** in cybernetics (a message that transmits information to an input about the output side) includes purpose, a tool that can compare the purpose and current performance and feedback process in various forms (Duffy, 1984). In the feedback process, there is an action that is carried out to achieve the goal, the result of this action, the effects of the result on the environment, the measurement of the difference between the result obtained and the goal, and thus a cyclical cybernetic model emerges (Dubberly & Pangaro, 2004) as depicted in in Figure 4.

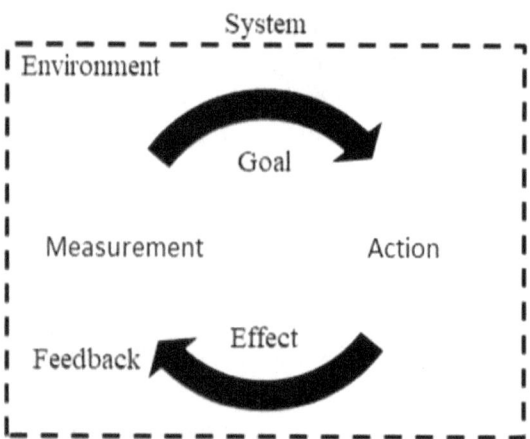

Figure 4: Cybernetics Feedback Model

The science of cybernetics focuses on the use of the **flow of information** as a tool for the self-management and control of the system with the feedback concept (Duffy, 1984). The control described here is not a restrictive and repressive control in the traditional approach, but tries to explain the limits of this independence of the units acting independently by the role of regulator (Turchin, 1999).

Beer (1968) answered the question of **how people can manage management** as in the form of knowledge and experience, and emphasized that the first output expected from managers is to produce policies by using the necessary knowledge. Beer (1968) emphasizes that it is not reliable for managers to produce policies on their own and mathematically proves that the method of ensuring high reliability is to obtain the necessary information from different sources[5].

However, depending on the level, how and by whom the information will be used, a filtration process should be performed and unnecessary information should be prevented to the levels that do not create added value. The filtration function can be explained by the fact that a physician develops a model primarily by using the basic indicators (fever, pulse, blood pressure, etc.) in the human body and applies further investigations if additional data are needed in the light of this model (Beer, 1974: 39).

Cybernetics predicts the need for new information triggering only a dramatic change and not to produce or demand unnecessary information in normal operation (Walker, 2001). Umpleby (2006) emphasizes that the production of information that does not create added value has no other role than to serve to increase the complexity that already exists.

Cybernetics attempts to understand systems using the features of viability, requisite variety and recursiveness.

Viability

The cybernetic approach focuses on the **viability of systems**, with the ultimate goal of designing new structures and ensuring the sustainability of existing structures (Rios, 2012). The viability is

---

5   The probability of a brain cell making a mistake is 1/200 and its reliability is known to be 0.995. If this cell attempts to take information from only two sources, each with a reliability of 0.70, it is more than half the probability of making a mistake (0.51). New condition; the probability that three cells (redundant structure) having the same degree of reliability (0.995) instead of one cell, receive information from two sources, each having 5 different information transmission channels and having the same degree of reliability (0.7), is likely to make one out of hundred million (P) = $[1- (1-0.3^5 (0.995)]^4 = 3.03 \times 10^{-9})$ (Beer, 1968).

defined as the maintenance of existence with a unique identity, and its existence should not be perceived as independent and disconnected from other beings in the environment (Beer, 1985: 1). Viable systems, in accordance with the mission of the organizations in which they produce output systems (Beer, 1985: 8)[6].

Cybernetics considers the most important factor for the viability of the systems as adaptation to the environment. It is known that the law of viability in complex biological organisms deals with the dynamic structures that allow the components of these organisms to connect in a harmonious way (Beer, 1985: 9). In order to achieve this harmony, it is seen that viable systems should be capable of making autonomous decisions (Espejo, 2003: 6).

Viable systems have their own mission statements and they have ultimate responsibility for their work in line with the policies and strategies of the whole they are part of (Walker, 2001). These systems also have the ability to establish, regulate, decide and function on their own (McEwan, 2001). Vore (1990) states that in order to talk about a viable system, it must first have a systematic aim and mission. It is of utmost importance that the systematic aim and mission are in harmony with the strategic goals and objectives of the whole system. Stafford Beer states that traditional approaches are inadequate for the harmony of the goals and objectives of the whole and the objectives of the sub-parts and that the problem of sub-optimization is inevitable (Vore, 1990).

In cybernetics science, it is accepted that the decision-making mechanisms of the main system in which they live, through learning, adaptation and adaptability of viable systems, are potential command and control backups (Whittaker, 2009). However, these systems should have a limit on their degree of independence (Rios, 2012) and should not jeopardize the viability and future of the main system they are part of (Walker, 2001). It would not be wrong to describe viable organizations as structures that are immune to diseases and adapt to environmental changes, as in biological organisms (Beer, 1985: 9).

---

6   For example; education and research that make a university a university. Education and research is a viable system and the faculties in which it is located and the departments in which the faculties are located are viable systems. Other support elements in the university (board, committee, senate, senate, library, dormitory, cafeteria, etc.) are not viable systems (Beer, 1985: 12).

In the science of cybernetics - in particular the VSM approach - the sustainability of the viability of the systems is realized through two basic mechanisms:

- Cohesion and
- Adaptation.

The cohesion mechanism aims to harmonize the objectives and interests of the independent units in the main system with the main system policies, strategies, objectives and interests and to develop the competitiveness of these units in the market (Golinelli, Pastore, Gatti, Massaroni, & Vagnani, 2002). It is emphasized in the research conducted by Espejo (2013) that it is very important to ensure full compliance in order to survive in an environment where there is a lot of complexity.

Adaptation mechanism is intended to have the capacity to adapt to changes in the external environment in time (Espejo, 2003: 17). Along with adaptation, learning, development and evolution are also considered to be one of the fundamental dynamics of the concept of viability (Beer, 1968). Beer (1974: 39) states that there can be no learning without defining the equilibrium state, that adaptation cannot be achieved without learning and that evolution cannot take place without adaptation and that the end of the organisms that fail to complete this chain will be like dinosaurs.

Requisite Variety

Ross Ashby's **Law of Requisite Variety** stipulates that a social structure must have the capacity to produce - at least - equal complexity that is the level of variety produced by the environment in which it is built, and thus turn into a viable system by effectively operating the adaptation mechanism. The variety that can be defined as the number of possible situations that a situation or problem creates or can create is used to explain the extent of a problem (Rios, 2012). The variety in question should be perceived as dynamic variety (Senge, 2011), not the variety of detail resulting from the ability to accommodate many variables, but the ability of the same cause or action to produce different results and effects in the short and long run. Variety can be created not only by the environment, but also by operational units that function independently within the organization. The flow of variety for an organization is shown in Figure 5.

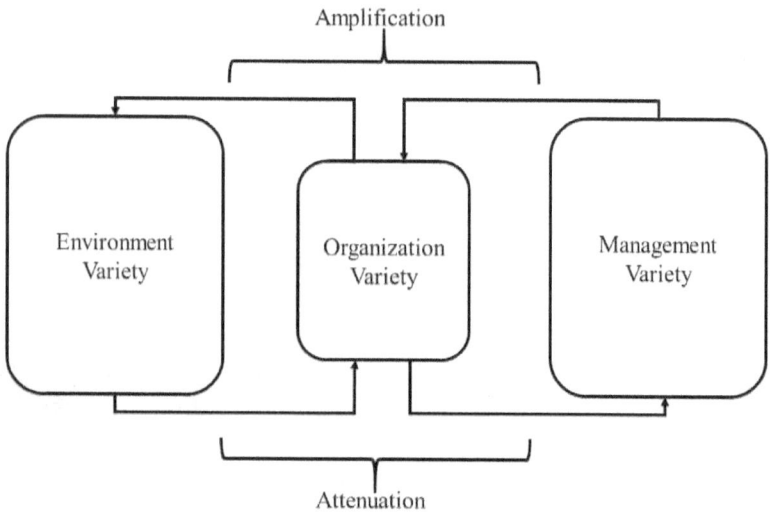

Figure 5: Variety Flow

While operational units are in charge of managing and absorbing variety from the environment, other subsystems within management have to manage variety arising from the organization (Walker, 2001; Rios, 2012). According to the Ross Ashby's Law, organizations use two main methods to demonstrate this skill. The first of these is factors such as authorization, individual initiative, production, training, employee engagement etc. that are used by the organization to amplify the variety produced and sent to the environment, the other is reporting, transparency, and being selective processes which can attenuate variety generated by the environment (Espejo, 2003: 10).

Amplification and attenuation methods, which are also defined as variety engineering, should be designed and used to serve the aims and objectives of organizations (Beckford, 1993: 16). Managers generally apply a rule-based (algorithmic) method to simple complexity within variety engineering, and heuristic or evolutionary method to high levels of complexity, and these choices are defined as their management techniques or styles (Vore, 1990).

Often used in cybernetics and VSM literature, Ashby's explanation of "only variety can absorb variety" (Walker, 2001; Whittaker, 2009) best summarizes the subject of variety. In this statement; in order to effectively manage a situation, it is stated that management units should have the ability to produce as much variety as the operational units they want to control (Beckford, 1993: 13). Beer (1994: 279), to explain the Ross Ashby's Law, that in a football match, two teams with the same number of players produce equal variety to each other, and if the referee toss a player from any team, the variety produced by the missing team is the variety produced by the opposing team. and that this team could be expected to lose the game.

Recursiveness

Recursiveness can be explained as the structure and functioning of viable subsystems that form a system and are intertwined, reflect or resemble the structure and operation of the system at a higher level (Beckford, 1993: 31). As in the matryoshka doll structure, there is an upper viable system in which each viable system is present (Beer, 1974: 17).

It is possible to see the projection of the recursion element in the VSM model as each operational unit includes sub-operational units and management units controlling these units (Beckford, 1993). Operations and management units defined at the lowest level should be perceived as Black Box.

The cybernetic definition of the term black box can be made in the form of a box in which the inputs and outputs are observed and the interaction between these inputs and outputs is unknown (Beer, 1994: 293). The Black Box technique focuses on the overall system, and foresees that the inputs and outputs are more concerned with the details of the sub-operational units (Beckford, 1993: 8). The philosophy behind the application of this technique is that it is assumed that the units defined as black boxes do not need to be opened to understand the functions (Beckford, 1993: 8). It can be said that the same technique can be used for the brain in biological structures and it would be a healthy approach to accept the brain as a black box by focusing on the functions of the brain rather than trying to fully understand the internal structure of the brain (Turchin, 1977: 118).

Cybernetics states that it is not possible or not necessary to fully understand complex systems, but managers can learn to organize inputs and classify outputs systematically (Beckford, 1993). In studies conducted with a traditional approach, the effort to collect information that does not create added value in understanding the whole system by opening the system parts that can be defined as black boxes is criticized by Beer (Vore, 1990).

In the operation of the Ross Ashby's Law; it can be said that recursive structure is a necessity for a lower level viable system to absorb the variety of a higher level viable system (McEwan, 2001).

VIABLE SYSTEMS MODEL (VSM)

Stafford Beer[7], a cybernetic scientist, defines the **whole** as the interaction of the functional parts that make up the whole within the **VSM approach**. Leonard (2009) states that VSM approach benefits from mathematics, psychology, biology, neuropsychology, communication theory, anthropology and philosophy of science. VSM developed by Stafford Beer can be defined as an approach to modeling an organizational structure based on the human nervous system (Umpleby, 2006). The schematic representation of VSM is as in Figure 6.

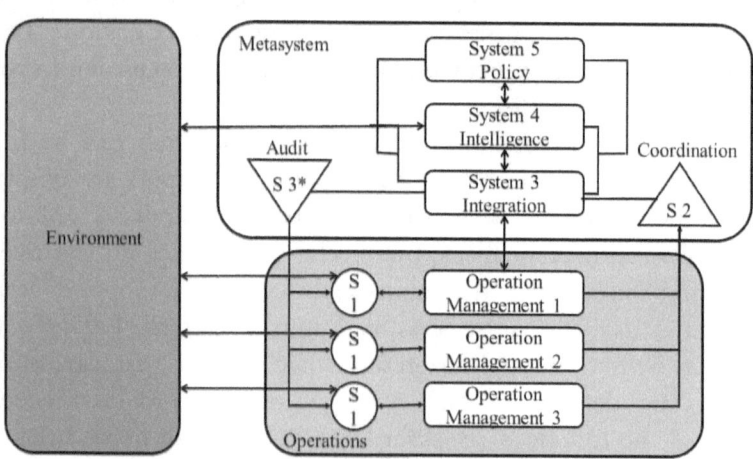

Figure 6: Viable Systems Model

---

7 Stafford Beer discusses application of cybernetic principles to organizational studies in his *"Cybernetics and Management (1959)"* book, use of cybernetic approach instead of mathematical models to solve problems in the field of management in his *"Decision and Control (1966)"* book, the differences in the management science in terms of system theory and management in his *"Management Science: The Business Use of Operational Research (1968)"* book, Viable Systems Model in his *"Brain of the Firm (1972)"* book, cybernetics science in his *"Platform for Change (1975)"* book, organizational cybernetics in his *"The Hearth of Enterprise (1979)"* book, application of VSM and cybernetics in Chile in his *"The Brain of the Firm (1981)"* book, application and schematic representation of VSM principles in his *"Diagnosing the System for Organizations (1985)"* book, and System-3 and System-4 of VSM and team syntegrity in his *"Beyond Dispute (1994)"* book (Rios, 2012).

Operational units, meta system and environment are included in the whole scope of the VSM. In VSM - necessary and sufficient (Espejo and Harnden, 1984) - operational units (System-1) and five subsystems in the meta system (System-2, 3, 3★, 4, 5) (Walker, 2001; Rios, 2012) are given in the following paragraphs.

Operational units are viable systems that operate the basic processes and produce outputs in line with the mission of the organization they are in (Beer, 1985: 8). The other units in the meta system have the role of establishing an environment that will enable the operational units to produce output in an integrated way (Walker, 2001).

System-1 (Operation): In order to produce the expected outputs; they are independent units that must perform basic functions and processes. The external environment is in relation to System-2, System-3 and 3★, indirectly with System-5 and other independent units. The outputs of a living system are produced by System-1, which can also be referred to as "system-in-focus". The independence of System-1 is governed by the rules previously agreed in the whole. System-1 can manage the changes and developments within its own structure with its independence. Independent units have their own missions and financial resources to realize this mission. The independence granted to System-1 also imposes the responsibility to comply with the policies and strategies in force for the whole system. Although all sub-systems in VSM function equally and in relation to each other, it can be said that System-1 is more specific than other subsystems. Because it is known that System-1 is the sub-system that includes all other systems due to its self-iteration feature. As a result, System-1 is the viable system itself (Walker, 2001: 8-17).

System-2 (Coordination): These are the regulatory mechanisms such as information systems, production plans, programming tools, processes, procedures to ensure the coordinated and integrated operation of System-1 and to prevent possible conflicts. In VSM, System-1 assumes the role of producing output, while the role of homeostasis is assumed by other subsystems. The whole system works to achieve this balance. However, the balance in the real world cannot be fully achieved and a continuous oscillation is observed within the system. This oscillation can be defined as the efforts of all elements in the system to balance their relations and adapt to the environment. System-2 is assigned to damp the oscillation. Since there are many oscillation sources in the system, multiple System-2 requirements are available for each System-1. The responsibility of determination and formation of the mechanisms to be used as System-2 is at the management level of System-1. Examples of typical System-2 mechanisms are timetables, production plans, personnel policies, information systems, coordination teams function as variety attenuators (Walker, 2001: 8-17).

System-3 (Integration, Adaptation, and Control): It enables System-1 to produce output in accordance with the determined strategies and policies. It allocates resources by focusing on creating synergy within the whole system. It also controls and evaluates efficiency and efficiency by collecting data via System-3★, a subsystem of System-3. System-3 is the mechanism by which independent System-1 interferes with the viability and future of the main system. The harmony between System-3 and System-1 is ensured through strategic planning. System-3, which deals with the management of the internal and daily functions of the whole system, plays the role of "here and now". System-3, which also has the role of operations director, observes and controls the relationships within System-1 and System-2. This system also fulfills the task of eliminating possible oscillations in System-2. System-3 uses System-2 and System-3★ to absorb the variety created within the whole system (Walker, 2001: 8-17).

System-4 (Information and Intelligence): These are the mechanisms that monitor, analyze and formulate strategic and operational decisions for the future in order to adapt to the external environment and to monitor the current situation and all changes (technology, threats, market, competition, export, etc.) that have occurred or are likely to occur. System-4 works as a mechanism for damping the oscillations that may occur in the adaptation of the whole system to the external environment. Unlike System-3, System-4 assumes the role of "outside and then" (Walker, 2001: 8-17).

System-5 (Policy): It is the highest level mechanism in which policies and strategies are determined, the interaction between System-3 and System-4 is managed (Umpleby, 2006) and an indirect relationship is established with System-1. This interaction of System-5 can be explained as balancing the harmony between System-3 and System-4. Since System-5 has a good command of the mission of the whole system, it is capable of absorbing the variety created by System-3. The indirect relationship between System-1 and System-5 is established by alerting System-5 with a signal (algedonic) sent by System-1 to convey distress or satisfaction in very special or emergency situations (Walker, 2001: 8-17).

VSM; in line with the principles of communication, self-management and control of the science of cybernetics (McEwan, 2001), tries to create a synergistic, flexible and creative system structure (Espejo, 2003: 4) by focusing on the interrelation of all the components of a system [even if these systems are in a bureaucratic organizational structure (Mabesa & Whittal, 2011)]. The relationship arises from the reciprocal interaction of people in the systems defined as a social structure. It will not be wrong to say that communication and process networks that enable people and other resources to work interactively are actually the organization itself (Espejo, 2003: 5).

The organizational structure envisaged by the VSM approach should be perceived as a functional structure (Veeke, 2003), not a hierarchical organization, but where the responsibilities in the systems are clearly defined (McEwan, 2001) in order to manage variety. Umpleby (2006) states that the awareness of employees about the functions of the organization will increase with this perception created by VSM.

Using the recursiveness feature, VSM tries to show that all organizations or systems are in fact similar to each other (each having operational units and a meta-system that allows these units to work in harmony and an environment in which they interact constantly) (Walker, 2001).

It can be said that VSM can be applied to any complex social community or organization where communication and interaction exist, regardless of job description and size. In terms of understanding and analyzing complex and large structures, the ability to treat them in the form of viable and similar sub-parts that can be analyzed and managed in a similar way makes VSM a powerful analysis tool (Walker, 2001). In addition, the fact that the basic features that make an organization viable is known is sufficient to say that other organizations with the same characteristics, regardless of their size, sector and purpose, have viability (Rios, 2012).

It will not be wrong to say that VSM is found to be a field of application and research subject in a wide range of different subjects and sectors. Stafford Beer's steel industry, textile production, shipbuilding industry, paper production, insurance industry, banking, transportation and training applications (Espejo & Harnden, 1984), Veeke's (2003) logistics model implementation, Henoch and Ulrich's (2000) logistics management system design application, Verdouw's (2010) supply chain model application, enterprise resource planning application by Badillo, Tejeida and Morales (2008) and service sector application of Juarez et al. (2010) are only some examples.

The most prominent of these applications is the study of modeling the Chilean social economy (Cybersyn Project – Cybernetics and Synergy) by Stafford Beer during 1970-1973 (Leonard, 2009).

In the literature, it is seen that researchers who are looking for alternative analysis methods frequently apply to VSM. For example; in a study on management control systems, it is emphasized that VSM is a very powerful tool for expanding the researcher's perspective (O'Grady, 2009). As a result of a study conducted by Schwaninger (2006) and examining five case studies; it has been shown that VSM is a very powerful tool for problem identification, data collection and analysis.

There are also criticisms of cybernetic approach and VSM in particular. Firstly, it is stated that VSM is a very complex tool and it is not possible to use it easily in problem definition and solution processes (Schwaninger, Rios, & Ambroz, 2004). Stating that Stafford Beer is trying to establish a hierarchical structure in organizations with VSM approach, Nechansky (2013) critises VSM in the framework of below items:

- The proposed five-system structure is not necessary for the viability of simple systems.
- It may be insufficient for complex systems.
- Beer does not constitute a viable system model and only poses cybernetic problems in organizations.
- The approach can be used in the process of creating concepts for organizations.
- VSM is incomplete and has limitations in explaining the viability of large organizations in particular.

Moreover; the following criticisms are also made to the YSM in the literature (Beckford, 1993):

- Under the heading of methodology; that it is not appropriate to base the VSM approach entirely on the comparison of mechanical and biological systems and that the variable of variety is not a measurable variable in scientific studies.
- Under the philosophy of science; that the cybernetic approach, which emphasizes equilibrium rather than change, is inadequate to understand the whole organization, that it cannot be realistic to operate in the light of predetermined goals in a dynamic environment, and that the cybernetic approach pushes individuals in the organization to the second plan.
- In the scope of the application; using cybernetic models in real life can reveal autocratic structures and it is very difficult to apply this approach.

Beckford (1993) states that the criticism of the complexity variable and the comparison of mechanical and biological systems in the literature are not widely accepted, but that additional research and interrogations can be made with particular emphasis on the philosophy of science.

Application of VSM

Systems thinking and cybernetic paradigm provide the opportunity to make an induction study without making generalizations that will eliminate real world complexity at the beginning of the system modeling study. In the formation of integrated models with the systems thinking, VSM, which is a very convenient tool to handle the systems as a whole together with their relations with the environment and their adaptation to this environment is used (Walker, 2001).

It can be said that VSM can be one of the most suitable methods if the system which is subject to modeling includes many of the features listed below:

- The complexity of the system is under consideration,
- The main actors and data sources within the scope of the subject are people and organizations,

- Adoption of induction approach,
- The obligation to evaluate the system as a whole and bring it to a level where it can be compared with other systems,
- The need for a functional measuring tool and,
- The goal of creating an integrated model.

In this context; the phases and steps for developing a VSM model in Table 2 can be considered as a guide.

Table 2: VSM Model Design Phases

Phase	Purpose	Steps
Preliminary Diagnosis	Describe the subsystems needed for viability and draw the VSM model	– Defining the system in focus – Determining the purpose of the system – Identification of system components – Drawing VSM model
Designing Autonomy	The establishment of the necessary independence and control mechanisms for the operational units to produce output	– Determination of the missions of the units – Planning the necessary resource – Establishment of control mechanism – Determination of autonomy limits – Determining the rules of intervention
Balancing the Internal Environment	The establishment of internal balance of the system by the relationship among System-1, System-2 and System-3	– Regulation of information flow between operational units – Development of information system – Raising the capacity of System-2 and 3

Phase	Purpose	Steps
Information Systems	To set up an information system that produces up-to-date information with performance parameters	– Analysis of existing information systems – Determination of performance parameters
Balance with the External Environment	Adaptation to the external environment by establishment of a System-4	– Determination of methods – Analysis of the current System-4 – System-3 and 4 relationship analysis – Ensuring traceability – Controlling resource sharing
Designing Policy Systems	The establishment of a mechanism in which the relationship between System-4 and System-3 is monitored and new policies are made	– Analysis of the current System-5 – Establishment of System-5 after analysis
The Whole System	The establishment of the whole system integrating all sub systems (System-1, 2, 3, 3★, 4, 5)	– Establishment of the whole system

Source: Walker (2001: 34-70)

In the integrated models designed, the responsible units produce output in accordance with the targets determined for the relevant system. The output of these units is possible by cohesion within the system and adaptation to the external environment. Each unit that produces output, within the scope of its autonomy, contains the elements of operation, management and environment just like the upper systems to which they belong. This is due to the recursiveness of the system.

Also, the viability of the systems makes each of them comparable to the other. VSM, which is a functional analysis tool that enables the systems to be handled in the same plane, provides this comparison.

REFERENCES

Ashby, R. W. (1957). *An introduction to cybernetics*. London: Chapman&Hall Ltd.

Badillo, I., Tejeida, R., & Morales, O. (2008). Viable systems model approach to enterprise resources planning systems. In *Proceedings of the 52ⁿᵈ Annual Meeting of the ISSS (International Society for the Systems Sciences)* (pp. 1-14). Madison.

Bartlett, G. (2001). Systemic thinking: A simple thinking technique for gaining systemic (situation-wide) focus. In *Breakthroughs 2001: Ninth International Conference on Thinking* (pp. 1-14). Auckland: Prodsol International.

Beckford, J. (1993). *The viable system model: A more adequate tool for practising management?* (Published PhD Thesis). Yorkshire: University of Hull.

Beer, S. (1968). *Management science: The business use of operational Research*. London: Aldus Books.

Beer, S. (1974). *Designing freedom*. Letchworth: Garden City Press.

Beer, S. (1985). *Diagnosing the system for organizations*. Chichester: John Wiley & Sons.

Beer, S. (1994). *Decision and control: The meaning of operational research and management cybernetics*. Chichester: John Wiley & Sons, Ltd.

Dubberly, H., & Pangaro, P. (2004). Introduction to cybernetics and the design of systems. Retrieved from https://www.pangaro.com/design-is/Cybernetics-minimized-v8b.pdf on 10.23.2019.

Duffy, P. R. (1984). Cybernetics. *International Journal of Business Communication, 21*, 33-41.

Espejo, R. (2003). The viable systems model: A briefing about organisational structure. *Systems Practice, 221.*

Espejo, R. (2013). Organisational cybernetics as a systemic paradigm: Lessons from the past – progress for the future. *Business Systems Review, 2*(2), 1-9.

Espejo, R., & Harnden, R. (1984). The viable system model: Interpretations and Applications of Stafford Beer's VSM. *Journal of the Operational Research Society, 35*, 7-26.

Golinelli, G., Pastore, A., Gatti, M., Massaroni, E., & Vagnani, G. (2002). The firm as a viable system: Managing inter-organizational relationships. *Sinergie, 58*(02), 65-98.

Henoch, J., & Ulrich, H. (2000). Agent-based management systems in logistics. In *Proceedings of the ECAI 2000 Workshop 13* (pp. 11-15). Berlin.

Hyötniemi, H. (2005). *Information and entropy in cybernetic systems.* Retrieved from https://pdfs.semanticscholar.org/a641/3d039d2e7fcd6 810a6c9d08d6692 c8eac3d4.pdf on 09.07.2019.

Juarez, A., Padilla, R., Pina, I., & Matamoros, O. (2010). Viable systems model and quality of hospitality services. In *Proceedings of the 54th Annual Meeting of the ISSS (International Society for the Systems Sciences)* (pp. 1-17). Waterloo.

Leonard, A. (2009). The viable system model and its application to complex organizations. *Journal of Systematic Practice and Action Research, 22*, 223-233.

Mabesa, J. M., & Whittal, J. (2011). Analysis of the current cadastral system in Lesotho using viable systems modeling (VSM). In *Africa GEO Conference* (pp. 1-16). Cape Town.

McEwan, M. A. (2001). Navigating complexity in organisations: An examination of the viable systems model. In *Founding Meeting of ECCON* (pp. 1-19). Netherlands: Lage Vuursche.

Nechansky, H. (2013). Issues of organizational cybernetics and viability beyond Beer's viable systems model. *International Journal of General Systems, 42*(8), 838-859.

O'Grady, W. (2009). An alternate management control framework: The viable system model (VSM). In *Auckland Regional Accounting Conference*.

Reynolds, M., & Holwell, S. (2010). Introducing systems approaches. In M. Reynolds, & S. Holwell (Eds.), *Systems approaches to managing change: A practical guide* (pp. 1-23). London: Springer.

Rios, P. (2010). Models of organizational cybernetics for diagnosis and design. *Kybernetes, 39*(9/10), 1529-1550.

Rios, P. (2012). *Design and diagnosis for sustainable organizations.* Berlin Heidelberg: Springer-Verlag.

Rosnay, J. (1979). *The Macroscope: A new world scientific system.* New York: Harper&Row.

Schwaninger, M. (2006). Design for viable organizations: The diagnostic power of the viable system model. *Kybernetes, 35*(7/8), 955-966.

Schwaninger, M., Rios, P., & Ambroz, K. (2004). System dynamics and cybernetics: A necessary synergy. In *International System Dynamics Conference* (pp. 1-19). Oxford.

Senge, P. M. (2011). *Beşinci disiplin* (Ayşegül İldeniz, & Ahmet Doğukan, Trans.). İstanbul: Yapı Kredi Yayınları.

Stewart, J. (2000). *Evolutions arrow.* Canberra: The Chapman Press.

Turchin, V. (1977). *The phenomenon of science: A cybernetic approach to human evolution.* New York: Columbia University Press.

Turchin, V. (1999). *A dialogue on metasystem transition.* The City College of New York.

Umpleby, S. (2006). *The viable system model.* Washington: The George Washington University.

Veeke, H. (2003). Interdisciplinary modeling for logistics design. In *Proceedings of the 10th European Concurrent Engineering Conference (ECEC 2003).* Plymouth.

Verdouw, C. (2010). *Business process modeling in demand- driven agri-food supply chains* (Published PhD Thesis). Wageningen: Wageningen University.

Vore, D. (1990). *A cybernetic analysis of the application of MIL-STD-1567A, work measurement to weapon systems acquisition management* (Published PhD Thesis). Ohio: The Air Force Institute of Technology.

Walker, J. (2001). The viable system model. A guide for co-operatives and federations. *SMSE strategic management in the social economy training program* [Online].

Whittaker, D. (2009). *Think before you think: Social complexity and knowledge of knowing.* Oxon: Wavestone Press.

ABOUT AUTHOR(S)

MEHMET HILMI OZDEMIR who received his BS degree in Industrial Engineering from Turkish Naval Academy (1994) and his Ms degree in Material Logistic Support Management from Naval Postgraduate School, Monterey / USA (2000) completed his PhD in Defense Management at the Institute of Defense Sciences of the Military Academy in 2015. Dr. Ozdemir who also served as program manager at NATO is still working at STM Defense Technologies Engineering and Technology Inc (STM). He is the manager of STM ThinkTech (Future Technology Institute) which is the Turkey's first technology-based think tank. His research and study areas are; systems thinking, cybernetics, complexity theory, modeling and simulation, defense management and performance-based logistics.

5

SYSTEM DYNAMICS APPROACH

YAVUZ ERCIL AND CIGDEM BASKICI

> *"...feedback systems [...] appear to provide a structure [...] that is equally useful even as one's attention ranges as widely as production, economics, psychology, or growth processes..."*
> – Jay W. Forrester

ABSTRACT

In this section, system dynamics models are examined. System Dynamics is a branch of Management Science dealing with the dynamism and controllability of systems. In this approach, he argues that modeling can be done on how the decisions made in the turbulence environment can affect the environment and then decision makers.

With such a dynamic modeling, it will be possible to see both the possible reactions to changing environmental conditions and the ability of the possible consequences of that reaction to influence future decisions.

In the chapter, some system dynamics concepts are introduced in the first title. These concepts (order, feedback, nonlinearity, and loops) express how to define a structure with system dynamics.

In the second title, the modeling process with system dynamics is introduced. In this process, starting from conceptualization and ending with validity, modeling stages are examined.

In the third title of the chapter, modeling tools are introduced. It is possible to model all kinds of structures using these tools. This model also describes these modeling approaches.

System Dynamics

The analysis methods used in the business world in the 1940s were mostly based on an understanding of mathematical and statistical linear models. During the World War II these methods gained a certain place among countries' efforts to overcome their rivals and were considerably improved. And after the war, they continued to be used as part of the economic efforts that quickly developed once again. For example, after the World War II, in oil refineries products were obtained with a minimum cost from the crude oil at the entrance by using linear programming in the production planning phase. Moreover, solution models for various problems in many management science practices were developed with this understanding.

These intense practices showed that linear models were capable of and sensitive to dealing with only a limited type of problems. As the number of variables comprising the problems increased, as the relationships among these variables complicated and as the interaction among the variables become circular in such a way that it encompassed the whole system, the function of linear models started to be insufficient, as well.

One of the first scientists who noticed this issue and tried to find a solution to this problem was Jay W. Forrester. Forrester was a professor at the Massachusetts Institute of Technology (MIT) Sloan School of Management in the late 1950s. While he was searching for the understanding that would eliminate the insufficiencies of linear models in the field of management, he named his first efforts "industrial dynamics". Industrial dynamics aimed to overcome the restrictions posed by traditional linear approach through the tools of the systems science. The most important one among these tools was **feedback**. The feedback approach, which systems scientists had worked on for many years, was quite useful for understanding nonlinear structures.

Another systems tool that Forrester's industrial dynamics approach utilized was **loops**. Loops were included in the approach as ordered meanings formed by a series of processes. The industrial dynamics which were based on these tools were used for solution of problems in work evaluation, identification of organizational and social relationships and their structural changes. When the approach was supported and developed with computer systems simulation, it quickly spread and improved. And as a result of this improvement, the name of the approach became system dynamics. Since the early 1960s when it first emerged, System Dynamics (Forrester, 1961) has been applied to a wide variety of problems in both public and private sectors. Large corporations and public agencies draw on approaches obtained from system dynamics models in policy and strategy designs and tactical and functional decision-making processes.

Basic Concepts of System Dynamics

In terms of System Dynamics, presenting how some basic concepts are defined will make it easier to understand the following explanations. In System Dynamics approach, the first concept to agree on is no doubt the concept of system. Many definitions have been made for explaining the concept of system (Boulding, 1956; Coyle, 1997; Forrester, 1961; Klir, 2013; Langefors, 1995; Miller & Miller, 1995; van Gigch, 1993; von Bertalanffy, 1956).

Although there are certain distinctions among these definition efforts, the common points below are remarkable. The system;

1. Concerns with entities: For the system, entity may be a physical, social or conceptual unit or variable.
2. Involves interaction: The system examines the interaction of entities with each other. The subject of this interaction may be communication (Langefors, 1995: 55), cooperation (van Gigch, 1993), or assignment (Klir, 2013), depending on the abilities of the entities.
3. The goal represents entirety: In order to talk about a system, entities have to gather as an entirety around a common goal. In this sense, the opposite of the concept of system is pile.

Within this framework, system may be defined as a **set of entities continuously and mutually interacting with each other to create an entirety that represents a common goal**. In this definition, the fact that entities continuously and mutually interact on the dimension of time is an important phenomenon to be stressed out. This interaction refers to causality inquiry for each entity. In other words, relationship, e.g. as in network science, is a qualitative inquiry rather than a quantitative value. Therefore, it is directional, and it is accepted that primary entity triggers relationships temporally, as well (Coyle, 1997: 18-24). Thus, the whole of the interaction defines a behavior. It is the concern of system Dynamics to analyze this dynamic feature. In another saying, system dynamics aims to analyze system behaviors (Senge & Forrester, 1980). Relationships and bonds among the entities of the system are characterized as the structure of the system (Lane, 2008). Within this respect, e.g. the structure of a system about an environment is defined with animal population, birth and death rates, food amount and mutual bonds among special entities or variables belonging to the system about a certain environment. The structure of the ecological system incorporates the variables affecting the system.

The logical basis of the System Dynamics theory is the understanding that the elements that actually prompt structures are simple movements in variables, and the situations that are thought to be complicated emerge due to bonds among variables (Grüne-Yanoff & Weirich, 2010). Interaction of all system variables effective on any structure creates complication. In the system dynamics approach, it is accepted that this complication can be understood through analyzing feedback structures, which are inherent in every relationship (Yang, Chen, & Yau, 2002). Therefore, it might be thought that feedback structures play a key role in understanding complication.

Each system's behavior is idiosyncratic, free dependent on the starting point and nonlinear (True & Asmund, 2002). The system realizes its behavior under the influence of its own structure and elements. The behavior is under the influence and within the boundaries of the structure. It can be thus argued that the structure of a system determines the behavior of that system. System dynamics is used to try to understand how the structure of a physical, biological or literary system leads to displayed behavior of that system. For example, by defining the structure of an ecological system, it might be possible to analyze the behavior temporally with system dynamics. The power of the System Dynamics approach stems from its ability to define the structure. Since – in the system dynamics approach – the structure can be defined not just physically but also conceptually, conceptual definitions such as motivation, tiredness and respect can also be included in the model. Thus, the model can create simulations similar to behaviors seen in the real world.

System structure

The philosophy of System Dynamics is founded upon the model structure that supports the goal of prediction and control (Flood & Jackson, 1991: 63-65). This understanding identifies the basic starting point in the formation of System Dynamics structure. There are four important features to be considered in examining the system structure through System Dynamics. System Dynamics models are expected to possess these features. They are;

1. Order,
2. Feedback,
3. Nonlinearity and,
4. Loops.

Order

Order, which is the key point in the development of the System Dynamics model, is about the number of accumulations that represent the structure. The accumulations determine the amount of values (time, information, raw material, labor, etc.) within the organization. Relationships and number of accumulations within the organization determine the order within the system.

Feedback

Feedback takes place between an element and another element directly related to it, or element series indirectly related to each other and the element that starts the motion. The process of personnel recruitment to the organization is related to the accumulation of present personnel number as a directly connected element. This relationship can accelerate or decelerate the recruitment process. This interaction among loops can be negative (decreasing) or positive (increasing). Negative feedback is an inhibiting and controlling effect. Positive feedback starts and increases both growth and weakening (Barlas, 2007; Morecroft, 1982). Positive and negative feedbacks are the focus of structure analysis (Figure 1).

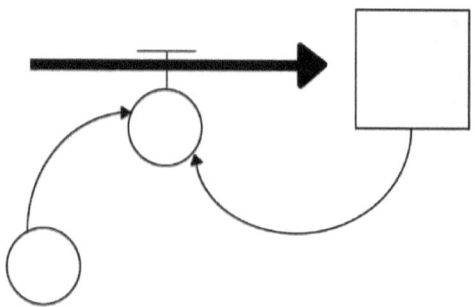

Figure 1: Feedback
Source: Developed from Richardson (2011: 226)

Nonlinearity

Systems influenced by positive feedback display an exponential improvement or regression at a determined point. Since the displacement between positive and negative feedback loops can be controlled, positive feedback – unlike linear systems – does not cause unnecessary harm to nonlinear systems. Thus, a controlled improvement takes place (Ajjarapu & Lee, 1992).

Loops

In reality, managerial, economic and social structures can rarely be defined by sufficiently simple loops. A few positive and negative loops get together and make a common definition. The number and degree of the mutual impacts between these loops create difficulties in defining key variables and predicting the outcomes. Therefore, it is hard to predict the outcomes without computer simulation. Without the help of simulation and analyses, behaviors emerge intuitively. It is accepted that, by putting forth solid causes and understanding the structure behind the development of the System Dynamics model, it is possible to provide high quality prediction and control.

Connection

Connection defines the interactions among the entities that constitute the system. The interaction among entities is the expression of the causality between the source entity and the target entity and creates **feed** on the target entity. Feed occurs in two ways, forward and backward. Feedforward is concerned with the system's connection to entities in its external environment. Therefore, the type of connection that is focused on while determining policies is feedforward. Feedforward is the opposite of feedback, which keeps the internal structure of the system stable. Feedforward is based on the prediction resulting from the system's reaction to stimuli and events and structural changes that can be applied or avoided in the future, are desirable or undesirable and disturbing. It carries out the reconstitution of the system structure and decision policies. Thus, desirable situations can be predicted (Wolstenholme, 2003).

SYSTEM DYNAMICS MODELING PROCESS

Even though there are different approaches regarding modeling with system dynamics (Boulding, 1956; Coyle, 1997; Worren, Moore, & Elliott, 2002), there are broadly five stages at the basis of modeling studies (Albin, 1997):

1. Conceptualization,
2. Formulization,
3. Test,
4. Implementation,
5. Validity.

Conceptualization

Conceptualization is the stage where the main frame of system dynamics is established (Figure 2). The goal with conceptualization is to concretize the subject to be studied. The main focus in conceptualizing System Dynamics should be the system structure. The first issue to consider while establishing the system structure is to determine the system's order. And the first step of forming the system order is to thoroughly define all entities that define the system. Variables that may affect the problem should be included in the model.

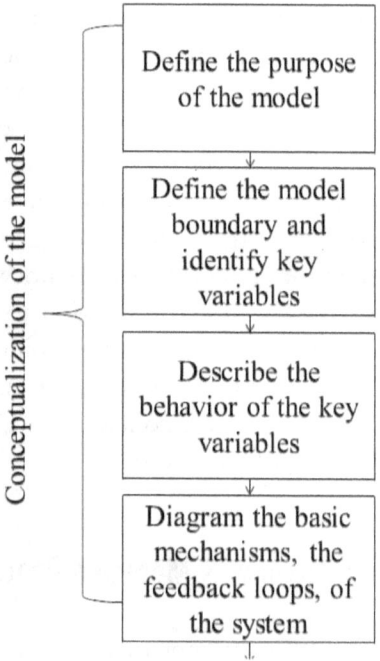

Figure 2: Conceptualization
Source: Extracted from Albin (1997)

Conceptualization is possible through definition within certain boundaries. The graphical marking approach, which is the causal loop diagram that determines the feedback's direction, can be used as auxiliary for this subject. Graphical marking is the expression of how elements interact with each other. After the system structure is explicitly ascertained through graphical marking, it is transformed into a flow scheme with mathematical expressions.

Formulization

Formulization is the revealing of the interaction of the relationships obtained through conceptualization. At this stage, component strategy can be used (Burns, Ulgen, & Beights, 1979). This approach includes the concept of an interaction matrix. The interaction matrix is a relationship matrix in which the relationships among the entities that constitute the system for model development. After the relationships in this relationship matrix are defined, impacts of these relationships are defined with formula. Another step at this stage is to set values for formulas.

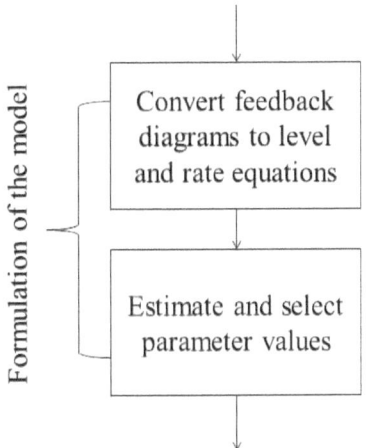

Figure 3: Formulization
Source: Extracted from Albin (1997)

In formulization, system goal is used for revealing important characteristics of the model (Blajer, 2001). After the model is defined and formulated as draft, it might need to be reexamined. At this stage, key subjects to be considered for making the final formulation are the availability of the data, related important theories, and scientific observations made for explaining events. At the same time, a big prediction series will be created with any modeling effort. These predictions must be put forth for increasing the explicitness of the model. The number and importance of these predictions will be associated with the existence, quality, theories and laws of data. Coherent formulization will facilitate the use of system dynamics as laboratory (Arango, Castañeda Acevedo, & Olaya, 2012).

Test

At the test stage, the structure of the model developed, the behavior and policy applications it produces are tested (Senge & Forrester, 1980). During this process, the first step is to test the dynamic hypotheses. Following this, acceptances of the model are tested and then the sensitivity of the model's behavior to difficulties is tested (Figure 4). The test tools according to Senge and Forrester (Senge & Forrester, 1980) are listed in Table 1.

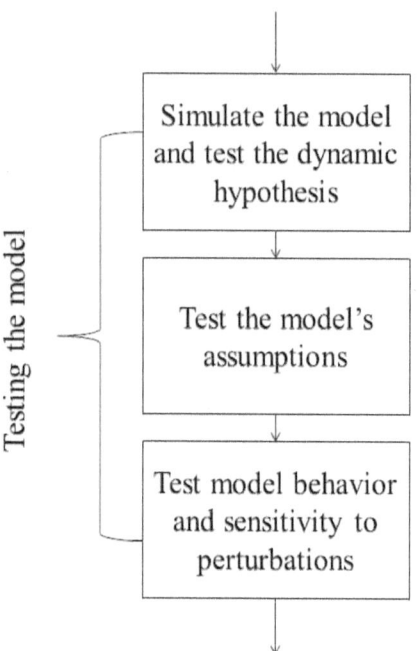

Figure 4: Phases of Testing the Model
Source: Extracted from Albin (1997)

Table 1: Test Tools for the System Dynamics
Model Structure and Behavior

Test Purpose	Test Tool
Testing the model structure	Structure verification
	Parameter verification
	Extreme conditions test
	Boundary adequacy test
Testing the model behavior	Behavior renewal
	Behavior prediction
	Behavior abnormalizing
	Family member
	Surprise behavior
	Extreme policy
	Boundary adequacy
	Behavior sensitivity

Source: Senge & Forrester (1980)

Implementation

The last stage of modeling is implementation (Figure 5). The first step at the implementation stage is testing the model for different policy implementations. The test tools suggested by Senge and Forrester (Senge & Forrester, 1980) for this stage are presented in Table 2. The second step of the implementation stage is documenting all modeling work information. This will help the study be evaluated by other scientists, as well.

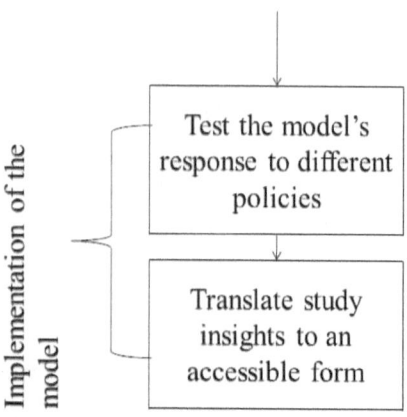

Figure 5: Implementation of the model
Source: Extracted from Albin (1997)

Table 2: Test Tools for System Dynamics Policy Implementations

Test Purpose	Test Tool
Testing policy implementations	System development
	Altered behavior prediction
	Boundary adequacy
	Policy sensitivity

Source: Senge & Forrester (1980)

Validity

After the implementation stage, the model's validity is required to be examined. Validation must be the most important and explicit part of the modeling stage (Barlas, 1996). In addition, it is very important to elaborate the validation procedure. The elaboration in validation must be conducted in various fields. The diversifications of validity are experimental validation, theoretical validation and pragmatical validation (Martis, 2006).

Experimental validation is carried out by questioning whether or not the results obtained by trying different values regarding the created model's variables meet the expectations. Experimental validation uses numerical and nonnumerical tests and sensitivity analysis (Lamperti, 2018).

Theoretical validation is carried out by comparing theories available in the scientific literature. The model does not need to comply with the generally accepted theories, but the comparison might prove the meaning and importance of the model (Mora, Cervantes-Pérez, Gelman-Muravchik, & Gelman-Muravchik, 2011).

Pragmatical validation is synonymous with the constant assessment made by implementers. When adequate compatibility is provided among the modeled elements, the model can be used for prediction and control. After validation, the conceptualization of the model will be carried out. However, the tests at the validation stage will continue at this stage, as well (Worren, Moore, & Elliott, 2002).

MODELING TOOLS

Realization of modeling processes depends on the effective usage of system dynamics tools. Basically, there are three system dynamics tools:

1. Accumulation,
2. Flow, and
3. Converter.

It depends on the use of these three tools to better define any model physically in system dynamics.

Accumulation

The first step is to define accumulations. Accumulation can be presented as a simple box (Figure 6). Accumulation functions quite like an accumulator and accumulates everything that flows within the system (Forrester, 1961). The food in our stomach, the money in our pocket, love in our heart, they are all accumulations. These fluctuate, end, increase and decrease. Accumulations imply both material and spiritual magnitudes. It is possible to understand the conditions of a system's elements by looking at the levels of accumulations at a point and time.

Figure 6: Accumulation

Magnitudes from flows gather in accumulation. The necessary question for defining accumulations is: What is the agglomeration in the system? For example, the most important accumulation for a work is workload.

Flow

Accumulations resemble pools, and just like all pool problems, there are faucets filling and draining the pool. The most appropriate question for defining the **faucets** of an accumulation is: What causes the workload accumulation? This question seeks answer to which tangible or intangible concept is filled into or drained from the pool, which we define as the **workload**. Thus, it is required to define flow in the second step.

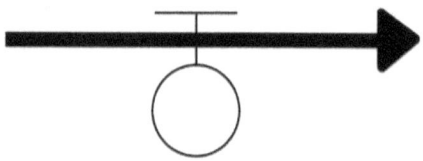

Figure 7: Flows

Within the system, physical or nonphysical entities move in channels called flows. Flows are displayed with a pipe, a faucet, flow adjuster and a one- or two-directional arrow. Figure 7 defines these constituents. The substance flows through the pipe in the direction shown by the arrow. These flows can flow into a void if they are not towards an accumulation. For system dynamics, void refers to the outside of the boundaries of the model structure, and is shown with cloud (Figure 8).

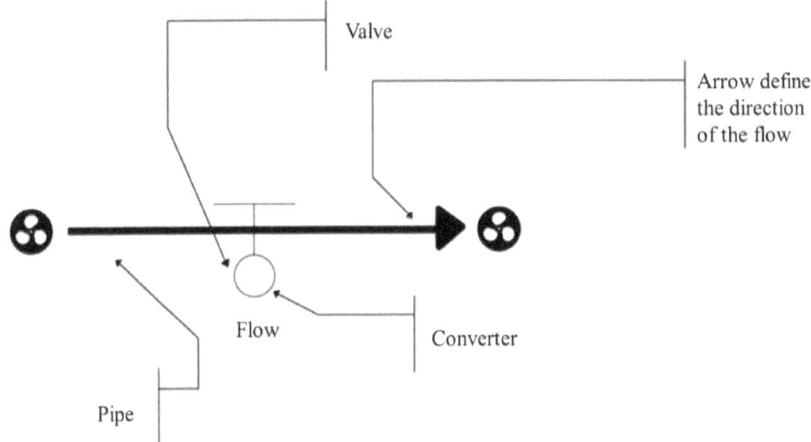

Figure 8: Parts of the Flow Scheme

The flow creates the circulation within the system and gives information about how the system operates. Physical or nonphysical magnitudes moving around in the flow gather in accumulations. In other words, there is no accumulation without flow. If there is an accumulation of something, it has to form as a result of a flow, an activity. And if there is a flow, there must be an increase or depletion. It is possible to establish the system model as a combination of these flows and accumulations. For instance, it is clear that the works carried out in a project leads to the increase of the workload. Since new works to be done emerge every day, works done do not have a permanent impact on reducing the workload. This simple system model's demonstration with system dynamics is presented in Figure 9.

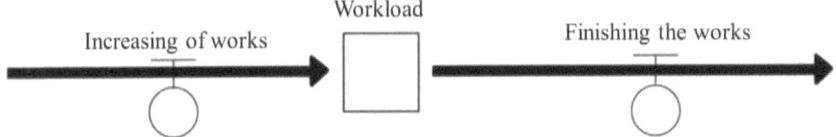

Figure 9: Change of the Workload

Within many systems, flows rarely provide a perfect balance. Think of an economic system. Producers rarely produce the same amount of goods as the consumers consume. Accumulations mostly function as a buffer within a model. It fulfils a **fill and drain** with flows and exits to maintain balance with another one. This buffering feature displays dynamic behavior patterns. Figure 10 shows these two flows and the accumulation between them called stock.

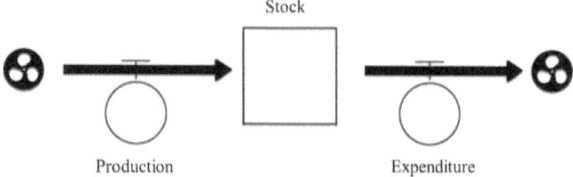

Figure 10: Buffer Effect of Accumulations

Accumulations in a system have a certain level even if all activities decrease to zero. If you take an instant photograph of a system that has ongoing activity, you see all operations frozen. In this picture, you only see the fields of accumulation obtained from activities. This picture shows you the condition of the system at a certain point at a time.

For example, consider that you stop all flows in a production process. What you can do is only observe the materials that seem to be accumulated in the system at that moment. This is partly demonstrated in Figure 11. Instead of the material flow within the combination process, you see the total amount of materials and labor within the system in different stages.

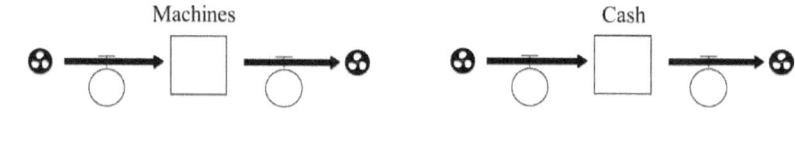

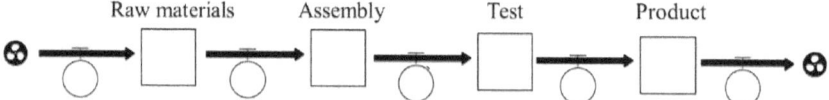

Figure 11: Combination of Flows

In terms of human resources, only the amounts are seen, not the motivation and morale of the personnel, and it is called high or low. When you check a bank account, you see the balance between the two sides of the account at that moment, rather than the flow of the money. The formation in both situations can be demonstrated with an accumulation. And by looking at the amount of materials within accumulations you can make a situational prediction about the system.

Let's consider a less physical example now. Assume that you freeze all activities within the system established to include you, your life, your friends and your loved ones. Just as in the previous example of production, conditions will remain the same even if activities are frozen (Figure 12).

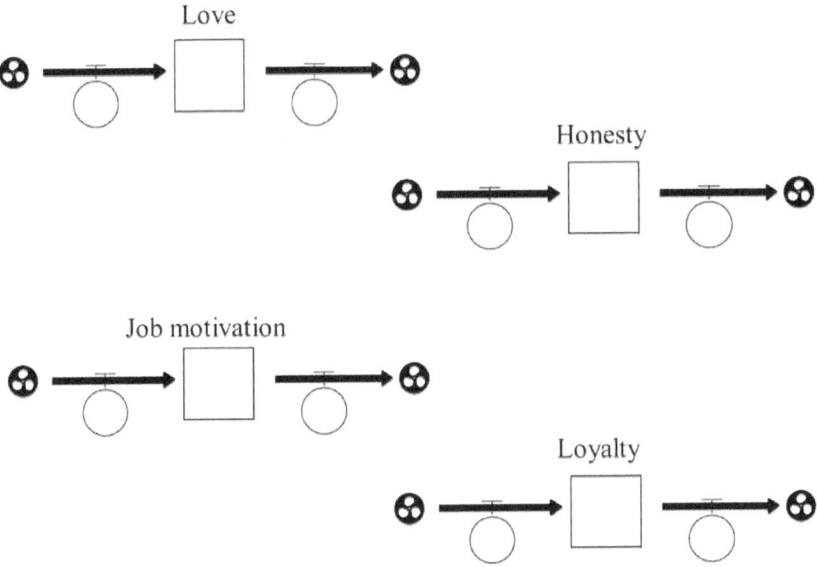

Figure 12: Demonstration of Abstract Concepts with Flows

The amount of love between you and your loved ones, the quality of relationship between you and your friends, your level of closeness to and disappointment with your job constitute the related accumulations. By looking at the levels of these accumulations, you can understand the level of your relationships with your friends and loved ones.

Converter

Converters usually serve the function of converting flows. But in the meantime, they are used for many other functions. They are signified by a circle (Figure 13). Converters convert inputs into outputs. They might indicate both information and material magnitudes. They are mostly used for defining the system logic that might be missed within the flow. They frequently include the magnitudes of variables (cost, loop time, etc.). Unlike accumulations, converters do not create accumulations. Values of converters are redetermined after every calculation. In this sense, converters have no memory.

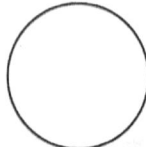

Figure 13: Converter

Connectors connect accumulations to converters, accumulations to flow adjusters, flow adjusters to other converters (Figure 14). Connectors represent inputs and outputs, not input flows and output flows! Connectors do not take up numerical values, they carry the values they take onto other elements. When you get onto the scales and check your weight, the numerical value on the indicator is carried onto your eyes. Such transfers are demonstrated through connectors. Connectors cannot be dependent on stocks. Only flows can change the stocks.

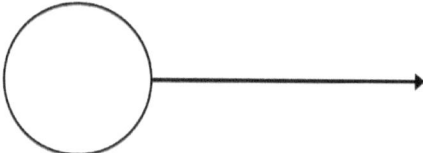

Figure 14: Connectors

FEEDBACK STRUCTURES

Assume that, the shape that will best represent the activities within the system for eliminating deviations from the goal in a project is to be determined by weekly productivity. Weekly productivity may be determined with hourly productivity and hours per week per personnel. Increasing or decreasing the hours per week per personnel can be interpreted as a function of workload and hours per personnel. In the meantime, it is possible to regulate the sub flow by using the information in the workload accumulation and thus adding feedback into the model. Because in this model, feedback determines workload and this is the balancing feedback loop. Feedback loop is shown within the revised model in Figure15 as a graphical function.

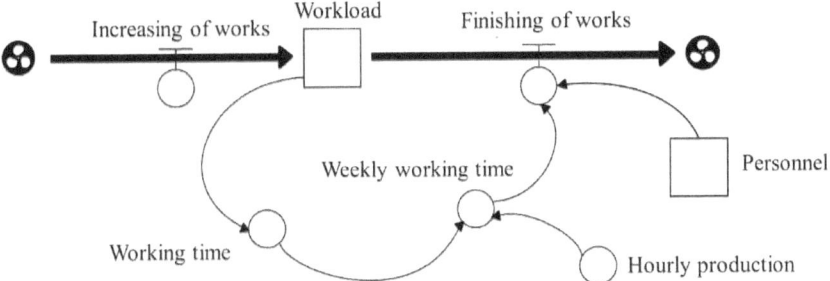

Figure 15: Improved Model

DIFFERENT DYNAMIC PROPERTIES OF ACCUMULATIONS AND FLOWS

Accumulations and flows, due to the nature of their definitions, have different dynamic properties. The differences between them may misguide people designing models for their system and lead to unplanned consequences through wrong revisions. For example, let's consider this report about population prepared in a growing country: **Despite the increase in female population, birthrate is rapidly decreasing, accompanied by very high mortality rates**. What would you think about the population status of this country? When considered superficially, it would be thought that the population is decreasing at a probably high rate. Because the differences in the nature of accumulations and flows work in the opposite way (Figure 16).

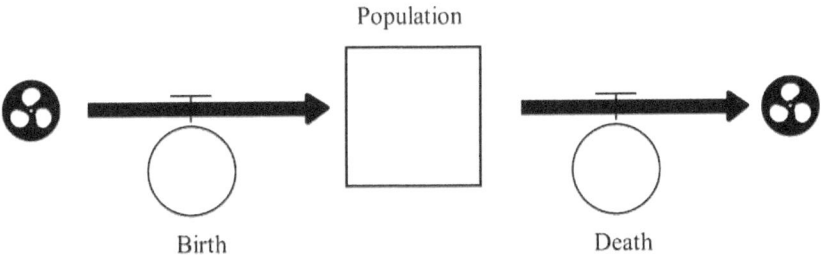

Figure 16: Simple Demonstration of Population, Births and Deaths

We can argue with a quick reasoning that population would decrease if birth rate was lower than death rate. Data about the decrease of births and increase of deaths are valuable information. Changes in the flow direction do not affect the changes in the stock direction. Only the conditions of the two flows with respect to each other determine the population accumulation level. When the number of births gets below the number of deaths, population decreases. Their rates do no matter. When we read the report, the situation in the flow immediately leads us to the easy conclusion we have first reached regarding the status of the system. However, those who have knowledge about how a system operates can differentiate between rate and situation, flow and accumulation.

For example, assume that we have found a vaccine for AIDS disease (extended from Sterman (2006)). Beyond that, let's assume that everyone in the world has been vaccinated. This means that no one will catch this disease any more (Figure 17).

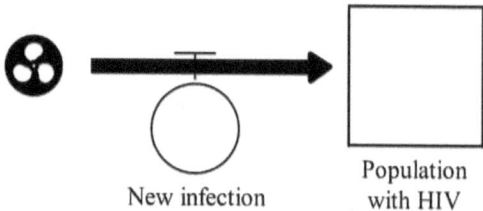

New infection   Population
                    with HIV

Figure 17: Affection by Disease

The flow that provides the input of newly infected ones will be completely zeroed. But unfortunately, the global population with AIDS still exists. Zeroing the flows that feed and empty an accumulation does not mean zeroing the accumulations, as well. Accumulation effect continues for a certain time. The change in accumulation will be equal to the net balance of input and output flows per unit time. This shows that accumulations indicate a system's history. The indications are usually flows. Because when the wheels of a car creak, people try to stop this creaking. But the real problem remains. The reason is mostly accumulations. We see once more that the difference between these two elements may lead to unproductive results.

A WORKPLACE MODEL

To reconsider the processes defined until now in an implementation, let's have a look at the modeling of the personnel working dynamics in a workplace as a generic event. The first thing to do is conceptualization. Let's first consider which entities (variables, for this subject) should be found for conceptualization (Table 3).

Table 3: Listing Basic Variables

Variable	Explanation
Work	Tasks assigned to personnel in the workplace
Workload	Accumulation of tasks
Personnel	People working in the workplace
Working	Effort made by personnel while doing work
Fatigue	Decrease in personnel's performance

Among these variables, workload, personnel and fatigue are dynamic concepts. That is, they are concepts that are expected to change over time according to the interaction of other variables. Therefore, they should be defined as accumulations. After this point, we can define the increase of works and completion of works, which are variables related to workload, as the workload's flows. Thus, our first flow-accumulation structure is established for modeling.

In the next step, the process above should be repeated for fatigue and personnel, as well. In this step, it is revealed that we have 3 flow-accumulation structures. The next is to include other variables that affect these flows and thus constitute the dynamic structure of accumulations in the model. To do this, we associate variables with flows and accumulations within their own groups (Table 4).

Table 4: Matrix for the Association of Variables

Variable	Work	Workload	Personnel	Working	Fatigue
Work		1	1	1	
Workload				1	

Personnel	-1				
Working		-1			1
Fatigue	1			-1	

After these relationships are obtained, it is possible to proceed to the formulization stage. For this, a mathematical evaluation is needed for every structure in which relationships are defined. For example, as the workload increases the need for personnel increases. This causality can be formulated as follows:

Personnel requirement = Workload * Workload increase rate

After all relationships are defined, a value prediction should be made for variables. At this stage, tests are done about which boundaries the variables should stay within and whether this will take place with formulas within the model. During the testing phase, tests are done for all three purposes (model's structure, model behavior and policy implementations). Multitude of test tools to be used here will enhance the model's resolution. After this stage, the last stage of implementation and theorization or policy analysis studies are done. The completed model of such a system is presented in Figure 18.

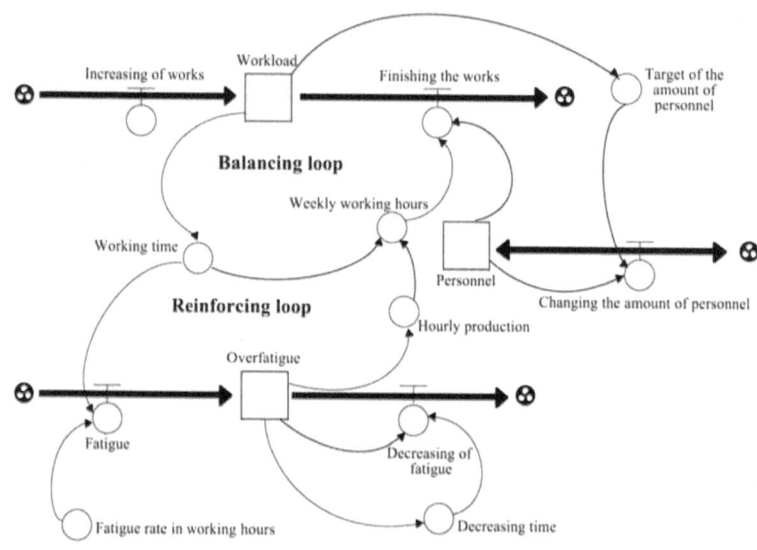

Figure 18: Completed Model

In conclusion, system dynamics uses computer simulation and various diagrams and demonstrations to model, assess and analyze system behavior. This approach makes it possible to conduct productive discussions on system structures with different perspectives. In this respect, system dynamics is a decision support system. Similarly, since it is an approach that enables certain experiments to be carried out on the system, system dynamics is also a scientific research method. In order to benefit effectively from system dynamics, the modeling process should be comprehended well, and modeling tools should be effectively used.

REFERENCES

Ajjarapu, V. A., & Lee, B. (1992). Bifurcation theory and its application to nonlinear dynamical phenomena in an electrical power system. *IEEE Transactions on Power Systems, 7*(1), 424-431.

Albin, S. (1997). Building a system dynamics model: Conceptualization. *Road Maps D4597,* 1-34.

Arango, A. S., Castañeda Acevedo, J. A., & Olaya, M. Y. (2012). Laboratory experiments in the system dynamics field. *System Dynamics Review, 28*(1), 94-106.

Barlas, Y. (1996). Formal aspects of model validity and validation in system dynamics. *System Dynamics Review, 12*(3), 183-210.

Barlas, Y. (2007). System dynamics: Systemic feedback modeling for policy analysis. *System, 1*(59), 1-29.

Blajer, W. (2001). A geometrical interpretation and uniform matrix formulation of multibody system dynamics. *Journal of Applied Mathematics and Mechanics, 81*(4), 247-259.

Boulding, K. E. (1956). General systems theory—the skeleton of science. *Management Science, 2*(3), 197-208.

Burns, J. R., Ulgen, O. M., & Beights, H. W. (1979). An algorithm for converting signed digrapls to Forrester schematics. *IEEE Transactions on Systems, Man, and Cybernetics, 9*(3), 115-124.

Coyle, R. G. (1997). *Management systems dynamics.* Chichester: Wiley.

Flood, R. L., & Jackson, M. C. (1991). *Critical systems thinking.* Chichester: Wiley.

Forrester, J. W. (1961). *Industrial dynamics.* Cambridge: MIT Press.

Grüne-Yanoff, T., & Weirich, P. (2010). The philosophy and epistemology of simulation: A review. *Simulation & Gaming, 4*(1), 20-50.

Klir, G. (2013). *Applied general systems research: Recent developments and trends (Vol. 5).* Springer.

Lamperti, F. (2018). Empirical validation of simulated models through the GSL-div: an illustrative application. *Journal of Economic Interaction and Coordination, 13*(1), 143-171.

Lane, D. C. (2008). The emergence and use of diagramming in system dynamics: A critical account. *Systems Research and Behavioral Science, 25*(1), 3-23.

Langefors, B. (1995). *Essays on infology: Summing up and planning for the future.* Lund: Studentlitteratur.

Martis, M. S. (2006). Validation of simulation based models: A theoretical outlook. *The Electronic Journal of Business Research Method, 4*(1), 39-46.

Miller, J. G., & Miller, J. L. (1995). Applications of living systems theory. *Systems Practice, 8*(1), 19-45.

Mora, M., Cervantes-Pérez, F., Gelman-Muravchik, O., & Forgionne, A. (2011). Modeling the strategic process of decision-making support systems implementations: A system dynamics approach review. *IEEE Transactions on Systems, Man, and Cybernetics, 42*(6), 899-912.

Morecroft, J. D. (1982). A critical review of diagramming tools for conceptualizing feedback system models. *Dynamica, 8*(1), 20-29.

Richardson, G. P. (2011). Reflections on the foundations of system dynamics. *System Dynamics Review, 27*(3), 219-243.

Senge, P. M., & Forrester, J. W. (1980). Tests for building confidence in system dynamics models. System dynamics. *TIMS Studies in Management Sciences,* (14), 209-228.

Sterman, J. D. (2006). Learning from evidence in a complex world. *American Journal of Public Health, 96*(3), 505-514.

True, H., & Asmund, R. (2002). The dynamics of a railway freight wagon wheelset with dry friction damping. *Vehicle System Dynamics, 38*(2), 149-163.

van Gigch, J. P. (1993). Metamodeling: The epistemology of system science. *Systems Practice, 6*(3), 251-258.

von Bertalanffy, L. (1956). General system theory. *General Systems, 1*(1), 11-17.

Wolstenholme, E. F. (2003). Towards the definition and use of a core set of archetypal structures in system dynamics. *System Dynamics Review, 19*(1), 7-26.

Worren, N. A., Moore, K., & Elliott, R. (2002). When theories become tools: Toward a framework for pragmatic validity. *Human Relations, 55*(10), 1227-1250.

Yang, S. K., Chen, C. L., & Yau, H. T. (2002). Control of chaos in Lorenz system. *Chaos, Solitons & Fractals, 13*(4), 767-780.

About Author(s)

Yavuz Ercil is professor in the Department of Public Relations and Publicity at Başkent University. He received his Ms degree from Istanbul University, Institute of Business Administration, Department of Management and Organization. His PhD degree was from Gazi University, Institute of Social Sciences, Department of Management and Organization. He works in the fields of strategic management, simulation, network science and system analysis. Since 2017, he is a member of the board of directors of Başkent University Center for Strategy and Technology.

Cigdem Baskici is assistant professor in the Departmant of Health Management at Başkent University. She completed her undergraduate education at Ankara University, Faculty of Political Sciences, Department of Economics. She received her Ms and PhD degrees from Ankara University, Institute of Social Sciences, Department of Business Administration. She works in the fields of international business, strategic management and network theory. Since 2017, she is a member of the board of directors of Başkent University Center for Strategy and Technology. She worked as a researcher in European Union and World Bank projects, and took part in projects carried out in cooperation with university–public institutions.

6

NIKLAS LUHMANN'S THEORY OF SOCIAL SYSTEMS

ELIF OZUZ DAGDELEN

> *"Reality is what one does not perceive*
> *when one perceives it."*
> *– Niklas Luhman*

ABSTRACT

Niklas Luhman is a prominent figure in the field of sociology that stands out with his social system theory. The social systems theory is conceptualized as autopoietic system theory that focuses self-producing process of society in the name of its own key elements and the boundary established as a result of this process. The emphasis of this theory is on the self-referent being of the autopoietic system. This system has inseparable parts like living systems, psychic systems and social systems and their mutual environments are important to comprehend the functioning. This functioning is explained with some special concepts such as operational closure, interactional openness, structural coupling, communication, function systems, interactions, organizations, codes and programs. Luhmann with these ideas and concepts tries to shed some light on the situation of modern societies. He discusses modern function systems as communicational organisms and evaluates them with functional differentiation, social subsystems, codes, functions, media, extra-social coupling, inter-social coupling, irritation, and resonance concepts. This chapter tries to explain the general frame of social systems theory and to describe the functioning of it with all of the concepts mentioned above. Thus, the development of the theory will be discussed and the contribution of the concept of love to the theory will be exemplified.

INTRODUCTION

Systems theories, after starting with biological base, find itself a place in other scientific areas too. Its effects and reflections are seen in sociology with structural functionalism especially with Talcot Parsons and Niklas Luhmann. However, through the socio-system theory of Luhmann, there are important developments in sociology. At the beginning, Comte's teological, metaphysical and positivist stages as lineer development; theories based on evolutionary and organismic views like Herbert Spencer provide basis for both structural functionalism and socio-system theory. In this chapter, the aim is to show the development of socio-system theory and its foundations, purviews and usages.

To do that, firstly what is the meaning of **system** in sociological thought and how **the thought of system** came in sight will be discussed. In this part, the dichotomy of structure and agency and its reflection on social system theory will be discussed. After that, we'll go over how system theory was developed in sociology, the understanding behind structural functionalism and meaning of it is explained especially with Talcot Parsons. Parsons might be accepted as one of the most important persons behind Luhmann's socio-system theory, therefore his scheme of AGIL, Theory of Action, ideas about social system and differentiation between inside and outside and important concepts like continued socialization, culturally structured shared symbols and internalization will be explained. Even if, Parsons is so important for understanding of Luhmann's theory, comparison between them will be done to show the differences between them and Luhmann's idiosyncratic qualities in the heading Talcot Parsons and Niklas Luhmann.

In the part of Niklas Luhmann and his **Social System Theory**, bio statement of Luhmann takes place. After that, to discuss socio-system theory with general system viewpoint, classical system theory, modern system theory and Luhmann's autopoietic system theory are compared and discussed in the name of their positioning of system and charachteristics of sytem. In relation with this comparison, the two person who tries to reconstruct Parsons' theory, in America Jeffrey Alexander and in Germany Niklas Luhmann will be compared. To relate Luhmann's with general system ideas, one of the pioneers of system theory Bertalanffy's ideas and the concept **autopoiesis** that can be seen as building stone of Luhmann's theory will be mentioned. After that directly, Luhmann's self-referential autopoietic system will be expounded.

The emphasis is to show that Luhmann sees society as autopoietic system. This system is self-referent. At that point, self-producing process of society in the name of its own key elements and the boundary established will be mentioned. As a kind of inseperable parts of autopoietic system, living systems, psychic systems and social system and their mutual environments will take place in this explanation. Luhmann's macro and micro understanding and concepts like operational closure, interactional openness, sructural coupling, communication, function systems, interactions, organizations, codes, programs will be clarified to see the functioning of society as a kind of social system or autopoietic self-referential system. After that detailed theoretical base, the situation of modern societies, Luhmann sees it as important will be expressed with the concepts like functional differentiation, social subsystems, codes, functions, media, extra-social coupling, inter-social structural coupling, irritation and resonance. It will be seen that modern function systems are actually communicational organisms and they have their own symbolically generalized communication media. This media has a diabolic function and it indicates and opens a road to new distinctions.

In the last part, **Luhmann's Social System Theory and Love**, we'll expand on how Luhmann tries to use these concepts and this functioning in the social sciences, to give example about a science of system in the name of him, love will be used as the main issue. The reason, why love is chosen to explain this is that Luhmann himself is very interested in this issue and has books like **Love: A Sketch"** and **"Love as Passion: The Codification of Intimacy**. In this chapter, love will be seen as a kind of medium of communication rather than individual feeling.

What Is the Meaning of "System" in Sociological Thought and How the "Thought of System" Came in Sight?

In sociology, the main dichotomy is between structure and agency or macro and micro. Because of this reason, to analyze society, the direction of the perspective matters as induction or deduction. It means that whether sociologists achieve knowledge through going beyond part to whole or whole to part matters in the name of sociological research. Especially in positivist epistemology, the reality is there and scientists are looking for reality in different manners because positivism foresees some kind of regularities in societies and these regularities give a general rule for the social system and in a different way supplies the continuity of society. **The father of sociology**, Comte might be a favorable person as an instance because he offers a linear development of society: teological, metaphysical and pozitivist stage. In this point of view, societies have some stages and this progress happens because of some social changes in society. Like all other natural sciences, society can be also analyzed in scientific ways and all of the problems of society can be solved thanks to science in that kind of view. To do this, societies should reach the positivist stage. Like Comte, first theories of sociology rely on positivist epistemology and ontology with the aim of finding regularities; solving problems and ensuring equilibrium.

However, to achieve this, if the methodological perspective is positivism, what is the appropriate perspective? To reach an answer, the question of whether structure controls individuals or individuals shape society takes place among all of sociological theories. However, at the beginning of the time sociology is becoming a science, to prove itself as a science, it is shaped through the methodologies and methods of natural sciences. Biology was one of the sciences that shapes the thinking of sociology with its evolutionary and organismic view. For example, Herbert Spencer, one of the pioneers of structural functionalism and social Darwinism, tries to make comparison between animals and humans. With regulative, sustaining and distribution systems he thinks that there is an evolution in both animals and humans. This evolution is also supported by the transformation of society with the rule of **survival of the fittest**. It explains that like in animals, in the society the powerful one takes the weak one. Spencer actually tries to understand the possible conflicts in society. With the organismic view of society that takes society as a body; institutions like the parts of this body and if there is any **disorder** (illness) in one of the parts, the equilibrium might be shaken. To the wellnes of body, all of the organs should work with their necessary functions.

The emphasis on function, interdependence of interdependent parts and seeing society as a kind of body or machine or system came in sight in this way in sociology. This kind of background formed the base of functionalism and specifically the base of structural functionalism. Even if functionalism appears as a counter argument for organismic view for criticizing it to take evolution as only factor for the societal change, organismic views then underlie the structural functionalist approach. The main aim of sociology for structural functionalists is to find the details of this process and to find the necessary conditions for integrity. Structural functionalists see individual as the core of society. Coexistance of individuals effectuates sub-systems of society and the functional aggregateness of these sub-systems gives rise to system. For structural functionalism, individual becomes **social actor** when he/she has a role and a status that is affiliated to this role. There are innumereus system theorists in the science environment but for sociology for structural functionalist like Parsons it can be seen his known book **Social System** is significant to understand what social system is and how society can be seen as a system.

In this chapter, the aim is to understand Niklas Luhmann's perspective on Socio-System and to do that it is inevitable not to speak of Talcot Parsons, functionalism and structural functionalism. The reason of this is that, theory of Luhmann's starting point is the critique of Parsons' structural functionalism. The establishment of Luhmann's theory of social systems grounds on Parsons' theory of system; Malinowski's functionalist method and Bellah's evolutionist differentiation theory (Çelik, 2007: 55). Functionalism isn't only significant per procuration of method but is important because Luhmann tries to reconstruct functionalism even if there are serious critiques towards it, later it is named as neo-functionalism. In addition to all of these reasons, the aim of this chapter, as well as understanding Luhmanns's perspective, is to comprehend it with sociological point of view.

How "System Theory" Developed in Sociology?

Structural Functionalism and Talcot Parsons

Structural functionalism takes society to the center and tries to understand how elements of society can be functional and functionally effective for the sake of it. The reason of this aim is understandable when their continuous emphasis on social order and stability is taken into consideration. Its focus is generally on macro that means structures and institutions and at the end to achieve equilibrium. To priotize structural functionalism's attention, large macro social structures and institutions and their relationships with each other should be thought. It's not enough to look at that but in the name of individual, it looks upon the restrictive effects of these and their relationship should be considered.

Talcot Parsons is an important person in sociology in the name of systems theory. He might be thought as a pioneer in developing strategy to see society in systemic ways. In his **General Systems Theory** sociology differs from psychology as a science of social systems and the latter one as psychic one. He tries to account for structure and conflict and, in a general manner, the main discussion of sociology: dichotomy of structure and agency. According to Parsons, humans are psychic systems who are trying to comprehend other social systems like themselves (Probert, 2014). Parsons is trying to explain what these systems are from the sociological perspective and he tries to do sociology of system. Parsons, as a structural functionalist, believes that system in sociology emphasizes a structure and its part; and if there is any change in any part it effects the other parts of the structure. This kind of society signifies the system. For example, if there is any change in economic system, political system will be affected.

In his famous book **The Structure of Social Action** Parsons makes explanations about the **action** systems and put four functional necessity towards it. These are: Adaptation, Goal Attainment, Integration and Latency. According to him, function means the things when they are unified, the requirements of the system can be met. These functions are known as his scheme of AGIL. Adaptation refers the adaptation of system to its environment; goal attainment means to priotize system's objectives and achieve them; integrity infers the organization of the relations of the parts of the system and to manage the relationship between A (adaptation), G (goal attainment) and I (integrity); the last one, latency signifies to the supplies, continuity and renewal of the motives of individuals in the systems. When the action systems are thought, behavioral organism is the function for the adaptation; personality system is the function for the goal attainment; social system is responsible for integrity and cultural system functions as the latency by transferring norms and rules (Ritzer, 2014: 244). Through these concepts and functionalist scheme, Parsons defines social system like that:

> "A social system consists in a plurality of individual actors interacting with each other in a situation which has at least a physical or environmental aspect, actors who are motivated in terms of a tendency to the "optimization of gratification" and whose relation to situations, including each other, is defined and mediated in terms of a system of culturally structured and shared symbols" (Parsons, 2013: 5-6).

In that way, social system is a kind of interaction system. However, rather than focusing on the concept of interaction he takes statuses and roles to the center of his theory. Individuals have functions in the structure **status** and they act with this status **role**. Functioning of the system necessitates the socialization and internalization. This necessity is based on his general theory of action. System means a kind of boundary. This concept makes a differentiation between inside and outside. To be inside, people give importance to the **approval** in action systems. In social integration, social interests don't create problem in this way because their ends are identified through shared values and norms to be approved. Like in the citation above, he makes an emphasis on social system, motives of people are identified according to the cultural system. They, through their continued socialization, internalize society's norms and values and this process give way to the possibility of social integration. Therefore, for him social system actually refers to the culturally structured shared symbols. To functioning of the systems, the subsystems like economy, politics, law, security etc. exist and function for the system.

Talcot Parsons and Niklas Luhmann

Niklas Luhmann developes a sociological approach that combines Parsons' structural functionalist approach's elements with general system theory (Ritzer, 2014: 163). In Parsons' social system theory, the concept **structure** precedes the function and systems concepts and in the formation of system **actions** are important and it continues with the logic of structural-functional system. Luhmann adds the structural one to post-structural frame. Rather than actions, hopes and expectations takes place in his theory. And the emphasis isn't like as the structural-functional system but like functional-structural system (Çelik, 2007: 56).

Rather than trying to understand social system through individual and action, Luhmann focuses on communication and communicational action in this autopoietic self-referential systems, this will be explained in the next part. The idea that, individuals have no role in the social system theory but systems are established with communication is a big difference between Parsons and Luhmann. The **metaphor of home** is used to explain this idea like that if individuals are some kind of brick for home, so society, when the construction of home is completed, it is meaningless to take care of bricks but the important thing is what is going on in chambers of home. Chambers of home in this part presents the subsystems of society: economy, art, politics, health, education, law etc. In this way, to define society, what people are doing doesn't matter but what social systems do matters. An end of an individual cannot decide the meanings in the society but meanings that come to mean as choices in the system identifies the social system (Cankurtaran Öntaş & Akçay, 2014: 102-103).

As it seen, the action for Parsons and Luhmann is very different from each other. "The foundation for Luhmann's sociology was then is now the systems- their etical, quasi-cybernatic reformulation of structural functionalism, especially of Parsons" (Misgeld, 1994: 153). Through the development of society, people's experiences with complex technologies change. As an example, Luhmann says, "computers are an 'alternative' to the structural linkage of communication and consciousness. Both communication and consciousness find ways to interact with computers because they seem to have their ways as up to now only communication and consciousness, apart from a sometimes whimsical nature, see and med to have had" (Baecker, 2006: 6). However, the most important point here is the choices systems give people and the process of becoming **observed observer**. Even if societies change, the not changing thing in Luhmann's theory is the preference of communication as a key term rather than action on the contrary to Parsons. However, the contributions and effects of Parsons shouldn't be ignored when Luhmann's theory is thought.

Niklas Luhmann and His Social System Theory

System theories have different shapes in different scientific areas. However, for sociology one of the most important names about system theory is Niklas Luhmann. He was born on December 27, 1927 in Germany. He studied law and worked as a lawyer until 1960 in different public institutions. In 1960-61 he carried on studies with Talcot Parsons in Harvard University and he took associate professorship degree in 1966 and started to give Sociology lessons (Cankurtaran Öntaş & Akçay, 2014: 98). Fuchs esteems him like that: "Niklas Luhmann, Professor Emeritus at the University of Bielefeld, Germany, died on November 12, at the age of 70, after a long battle with cancer. We have lost one of the most distinguished scholars and sociologists of our time. At the core is a theory of social systems, understood as recursive networks of communication and observation" (1999: 117).

Systems theory in social sciences is about the structure of society, its organization and focuses on how it functions. In general manner system theories takes functionalism and organismic view as a base. As a social science, history is an important factor for Luhmann and he takes into consideration of the evolution of societies by time. His theory and his explanations about social system as autopoietic system is explained with the concept of functional differentiation. To see the establishment of functional differentiation, history is important because it happens as a result of the evolution of societies. Four important developments come in existence throughout the history: (1) segmentary differentiation, (2) center-periphery differentiation, (3) stratified differentiation, and (4) functional differentiation. The first one based on the social subsytems' equality so it means the living that has no center of social power like the societies which have descent or communal living. The only stage that has equality is actually the first one. Unlike this one, center-periphery differentiation is based on structural inequality and it has a kind of emphasis in the name of location like it can be remembered from Wallerstein's theory. In this one, there is one segment that is powerful. Inequality happens in the third one again but this time it doesn't based on location but based of rank. Example of this differentation can be Indian Caste System or Medieval Europe. The last one, functional differentiation, is based on the subsytems' equally unequality. Subsytems are distinct from each other and humans can't locate themselves to one social subsytems. The focus here is that it doesn't refer the disappearance of social ranks but refers that social order is non-equivalent with class distinctions (Moeller, 2006: 49). Luhmann's concept functional differentiation is an important concept to understand modern society in a democratic perspective. Through this concept, without any intervention the functioning of all sub-systems like politics, law, economy and etc. are tried to be shown.

The question is, with this evolutionary model, that what is these mentioned systems and its subsystems and how the theory of social system of Luhmann is different from other ones? As Ertong shows that (like in the Table 1) there are classical system theories, modern system theories and autopoietic system theory. In the first one positioning of the system is totally based on whole-part difference. It is also one of sociology's basic discussions. Like in functionalism, society is a kind of system it is like a whole, it has parts and these parts form the whole (society). In classical system theory it is closed sytem. In modern system theories, system takes shape as opened one because they think that there is an interaction between system and environment. In Autopoietic System Theory, it is associated with, Niklas Luhmann, there is a differentiation betwenn system and its environment, and the issue is here not about openness and closeness but about their co-existence.

Table 1*: Change of Paradigm in System Approach

	Classical System Theory	Modern System Theory	Autopoietic System Theory
Positioning of System	Whole–Part Difference	Interaction of System-Environment	Differentiation of System-Environment
Charachteristic of System	Closed System	Opened System	Co-existence of Openness and Closeness

* Table is translated in English from Turkish
Source: Ertong (2017: 9)

Even if the arguments against functionalism appears in a visible manner at that time, in America Jeffrey Alexander and in Germany Niklas Luhmann try to reconstruct Parsons' theory in a multidimensional way. However, while Alexander focuses on individuality, dynamics of group, contingency and voluntarism; Luhmann tries to sustain classical system perspective so he follows generalized evolutionary perspective and uses organismic metaphors (Özdemir, 2001: 12 as cited in Çelik, 2007: 55). The source of this organismic metaphors and evolutionary perspective is actually based on Ludwig von Bertalanffy' general systems theory. He makes an important contribution about analyzing human being. Rather than taking them as robots, he created a new image for them. According to him, "human behaviour couldn't be explained purely in terms of biology, that is also influenced by the individual's understanding of self in relation to the world. Human beings are creatures of two worlds, biological organisms living a universe of symbols" (Hammond, 2010: 106). He refers the systems of values at that point. His theory can be seen as the system theory of organism. Living organism can not be analyzed in an isolated way. The forces that is inside that organism should be understood and analyzed. In this way, mutual interactions of complex elements inside of the organism are important. However, inside isn't enough and the relationship between them should be considered too.

Bertalanffy's ideas about system affected Luhmann's ideas so much and he created a socio-system theory also by using other social system theorists' concepts. Luhmann defines system with the concept of **autopoiesis**. Autopoiesis comes from the Maturana and Varela (other system theorists who are neurophysiologists) and it is the **auto** (self) **poiesis** (production) so self production. Through this concept the differentiation between system and environment and the creation of the systems by itself is emphasized. Ritzer asserts that, according to Luhmann, society creates its own key elements as self-producing system; it forms its own boundaries and structures, it is based on self-reference and it is closed, the corner stone of society is communication and communication is created by society (2011: 197 as cited in Cankurtaran Öntaş & Akçay, 2014: 101).

Systems theories are generally based on the natural sciences and the aim is here is to start with the depiction of the production of cells. Biological cells in this understanding help to understand the world. At this juncture, it can be said that Luhmann sees society as an autopoietic system and this system should be self-referent and to be self-referent it should include living systems (cells, brains, organisms), pychic systems and social systems (function systems, organization and interactions). These are seen as **mutual environments** that mean each system is in the environment of the others.

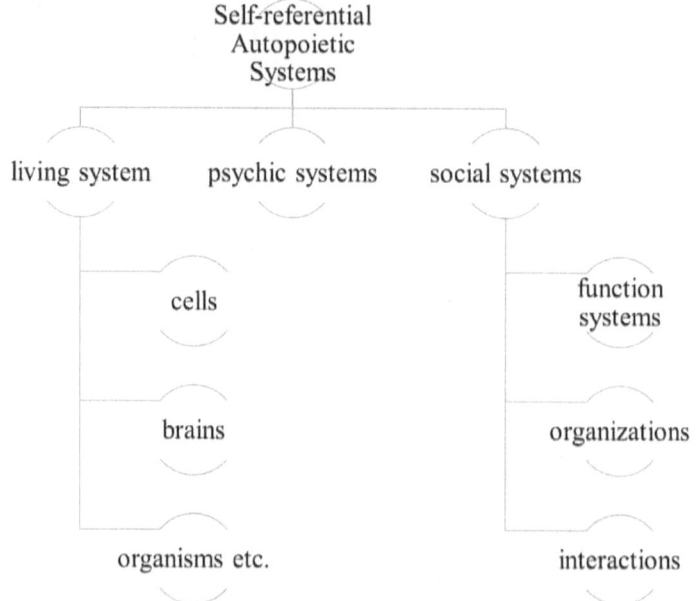

Figure 1: Types of Systems
Source: Moeller (2005: 11)

In that point, social systems are the systems that have their own environment but rather than direct action and reaction, their relationship is much more complex. The concept **mutual environment** doesn't mean that but on the contrary means that social systems draw a line between their environments and themselves. Luhmann thinks that this line should exist. It entails, distinction of inside and outside (like Parsons said but in a different manner), and system and the environment. Systems can not be understood without reference to their environments but the other way rounds to understand them the differences and distinction between them should be analyzed. Environments (the different ones) don't hinge on each other but connected with the system. Therefore, environments are always more complicated.

The concept environment is not only micro or macro. This macro and micro dichotomy is much more different than general sociology discussions. It is macro when Luhmann refers to the society to the world society and is micro when he means culture, policy, economy and family. Even the mentioned complexity exists around the social system Luhmann thinks it lives through decreasing that. While this process, he makes emphasis on that "autopoiesis entails: operational closure. The operations inside a system must relate to the other operations previously produced by the same system" (Guy, 2018: 292). Operational closure isn't closeness between systems directly. Operational closure in autopoeietic systems means continuous contact of system with its environment. It is again can be understood with biological concepts. For instance, living cells cannot exist without the excambium of energy and matter that they need. Unlike this, autopoietic system itself organizes the relationship with the environment and the decisions about energy or matter like the channels they will be exchanged through are made by the system (Seidl & Becker, 2006: 15). Therefore, systems like living systems, psychic ones and social systems are operationally closed but they are open to each other. Kieser examplify psychic systems like that: "As a psychic system, I develop ideas for and strategies on what kind of communication to offer the science system (this does not mean that I, as a person, am considered irrelevant, as some critics think; on the contrary, the two systems are relevant for each other and can be analyzed in their strategies toward each other)" (2007: 991). In this way, operational closure is seen as the first step for interactional openness.

The brain represents the living systems, the mind represents the psychic systems and the society represents the social systems. There are structural couplings between them. Structural coupling is the process in which one system shapes the other's environment. This creates and increases the mentioned structural complexity and to the continuation autopoiesis of the systems that is necessary. Luhmann explains that, "We can ... conclude that society only couples itself with its environment through consciousness, and that thus there are no pyhsical, chemical, and also no purely biological effects on social communication. Everything has to pass through that eye of the needle of communication" (2002a: 113 as cited in Moeller, 2018: 20).

It brings the most important concept **communication**, a kind of social operation, to light. Communication is a **self-observing operation** for Luhmann and like Habermas he thinks that one of the illustrative strains of social is communication. However, unlike Habermas, he doesn't see it as the necessary keystone to establish a consensus but sees as a necessary thing to "give rise to understanding, and as such is the control feature of all social systems" (Probert, 2014). It is not only understanding communication gives but also the unity of it with announcement and information. Luhmann thinks that self-referential autopoietik systems exist when communication develops from communication. For him: "Communication is the structural equivalent of biochemical statements by means of proteins and other chemical substances. It is of primary importance that there is a prospect of identifying an operator that makes possible all social systems, no matter how complex societies, interactions or organizations might become in the course of evolution" (Luhmann, 2006: 47). Society doesn't live directly but he uses biological concepts to explain it. Society can be seen as an organism but only as metaphorically. Society lives through communicating.

Society in that way is a system of communication. Like in the schematic categorization of self-referential autopoietik systems, systems can be categorized as systems of communication (social system), systems of information (bodies, brain and so on) and the systems of consciousness (minds). When society, a social system, is taken as large communication sytem and its autopoietic and self-referent structure gives the term autopoietik communication systems.

Autopoietic communication systems are based on function systems, interactions and organizations. The important emphasis is here that its basis is on difference. Communication is dependent on functions and function systems work with operationally closed systems like economy, politics, law and mass media as it is mentioned above. However, the functioning of the system there should be codes and programs. Programs can be thought as a kind of guidelines and codes as a kind of differentiations. For instance, when the law is taken as an example, constitution might be program and the code might be legal and illegal.

Luhmann, through function system, actually tries to talk about modern society's function systems. In modern societies, functional differentiation of society becomes involved. It means that if other systems cannot deal with societal issues or if the system isn't established, modern society can regulate itself by appointing various functions to specialized social systems. As it is mentioned above, modern societies come into existence at the end of the social evolution by following the segmented or stratified traditional societies that were structured hierarchically. Unlike these societies, spheres of society like economy, politics, science and law in modern one become differentiated from each other to fulfill different functions. For instance, in the sake of functioning of society, economy fulfills the need of distribution of outputs and services; politics are responsible for making necessary decisions; science procures the knowledge; law ensures the fairness, etc. It doesn't come to mean that these spheres like economy, politics, science and law etc. appear totally in modern society but it means that they become autonomous by carrying out their functioning (Görke & Scholl, 2006: 647).

"The functionally differentiated society consists of distinct functional subsystems that are specialized in serving specific societal functions; for example, law, science, economy, art, religion" (Seidl & Mormann, 2014: 19). These social subsystems also function with codes, functions and media. Social subsystems have spesicifis functions, efficacy, codes, programs and medium and that can be seen in Table 3. Structural coupling in that kind of systems are varied too like **extra social coupling** that is between communication and psychic systems and communication and **inter-social structural coupling** that is between continuous irritation and resonance relationship. In the first one, even the systems are operationally closed thanks to the interaction individual mind and society can come together. In the latter one, mentioned relationship, irritation and resonance, explains a kind of relational change and equilibrium between social systems like economy and politics.

Table 3: Social Systems

System	Function	Efficacy	Code	Program	Medium
Law	Elimination of the contingency of norm expectations	Regulation of conflicts	Legal/illegal	Laws, constitutions, etc.	Jurisdiction
Politics	Making collectively binding decisions possible	Practical application of collectively binding decisions	Government/ opposition	Programs of political parties, ideologies	Power
Science	Production of knowledge	Supply of knowledge	True/false	Theories, methods	Truth
Religion	Elimination of contingency	Spiritual and social services	Immanence/ transcendence	Holy scriptures, dogmas	Faith
Economy	Reduction of shortages	Satisfaction of needs	Payment/ nopayment	Budgets	Money

Resource: Krauze (1999: 36 as cited in Moeller, 2006: 29)

Modern function systems are considered as a kind of communicational organisms too. In that point, they have their own media and in these systems language become at one point senseless because each subsystem has its own medium. All of them have their symbolically generalized communication media. For example, medium of economy is money; medium of law is jurisdiction, medium of politics is power and the medium of science is scientific truth. Luhmann thought that, this media has diabolic function because of inclinations of new distinctions. There can be seen some examples where money become more important than faith or the reverse. It cannot be ignored that according to Luhmann, media face with an evolution too and there are two kinds of media as disseminating media and success media. Example for disseminating media can be television or e-mail and it can reach small group. However, through evolution and social differentiation the success media is developed. The term that is linked with modern function systems: symbolically generalized communication media is actually success media. "Success media are symbolically generalized media that convey meaning within a specific societal system. A specific binary code produces the meaningful form, difference, within each success medium" (Lee, 2000: 326).

The second autopoietic communication system is interactions (the first one was the function systems as it can be seen above). Interactions are important and different from others because there is physical presence of participating individuals. However, this isn't total physical existence but the reflections of it on communication. Luhmann explains interaction systems like that: "interaction systems include everything that can be treated as present are able, if need be, to decide who among those who happen to be present, is to be treated as present and who not" (1995a: 412 as cited in Seild & Mormann, 2015: 20).

Third type of social system is organizations. Organization signifies the systems of decisions. Seidl and Becker says that, "Luhmann's theory of social systems sees organizations as consisting of communications or- in the case of formalized organizations- decisions, i.e. actions under the pressure of expectations" (2005: 117). For example, an emergence of politics as an autopoeticsystem might signify an emergence of political parties or the differentiation of education system signify an emergence of educational organizations. Organization might also carry the meaning of membership. However, Luhmann asserts that they don't have to stuck with the border of one function system. On the contrary communicative borders can enlarge and compose each other. As an example, the multi-varied composition of universites can be showed. Universites functions as both educational and scientific areas and also can be considered as economical ones. There are again complexities in organizations. Luhmann talks about organizational life and the necessary elements to reduce this complexity. According to him, "Deadlines are structural means of reducing complexity. Organizational life is largely characterized by the priority of time-limited issues" (Luhmann, 1971b: 143). Therefore, deadlines are important for equilibrium. With this aim, he is interested in schedules and deadlines. "Luhmann examined how schedules and deadlines reduce the complexity of organizational life by determining work rhythms and the choice of topics. Such strategies are considered typical of organizations that attempt to cope with the complexity of the environment without either being overwhelmed by it or oversimplifying it. As a means of reducing complexity, deadlines filter facts and social coordination and make them manageable" (Seidl & Mormann, 2014: 8). Equilibrium in Luhmann's perspective doesn't mean the extinguishment of conflict. Luhmann, while he gives importance to the consensus, he never ignores conflicts. On the contrary, he sees it as **integrated systems**. The issue is here to deal with conflict by comtemplating loose coupling as a solution (Probert, 2014).

In conclusion, through functional systems, interactions and organizations, Luhmann actually achieve the idea of **World Society**, as mentioned above. Stichweh, asserts that by comprising all interactions and organizations, society symbolizes the largest social system. Likewise, Luhmann by naming today's uniqe and the largest social system as World Society, identifies it with communicative attainability (1999: 22).

LUHMANN'S SOCIAL SYSTEM THEORY AND "LOVE"

Luhmann's social systems are applicable to a lot of area. Its importance and usage are increasing in different scientific areas and concepts. For instance, Luhmann himself analyzes love as a kind of medium of communication (Love: A sketch and Love as Passion: The Codification of Intimacy) law as a social system (Law as a Social System) and religion as a social system (A Systems Theory of Religion). In this part as an example Luhmann's views on love should be showed.

To understand his social system theory love can be a good instance. In Love: A Sketch, about love, it is seen that rather than individual feelings, the institutionalization of love and socio-structural conditions take place. It is seen that individuals and their feelings or action don't matter. Love matters and exists because it is institutionalized because the feeling people think that it is an actual feeling, according to Luhmann are effects of cultural socialization. Like it is mentioned above, language doesn't matter because each subsystems developes its own medium. Luhmann thinks that, love is a kind of medium of communication.

By using the category 'medium of communication', it is also clear that we are not seeking to address love in this context as an objectively identifiable feeling of a particular kind or to determine its occurrences, provide casual reasons for them or render them functional in terms of individual's organic or psychological system. For our argument, the reverse is the case: a certain ambiguity and plasticity in emotional state is essential (although, of course, love as a medium of communication, is not compatible with each and every motivational structure). It may well be that the breakthrough leading to one's first taste of independence from one's parents, the excitement experienced during one's first tentative encounters or the first experience of mutual recognition with a sexual partner is, with the aid of cultural cliche, interpreted as love- and then turned into love (Luhmann, 2010).

Like autopoietic communication systems work with function systems, interactions and organization; love can be understood with these concepts. The necessity of functioning of that is also based on codes and programs. Functional spesification or differentiation of love can be seen by history again. Cultural socialization in this medium is so important and much more important than individual feelings. Love isn't different from money, power or art. The code for love is then **being appropriate or not** (Luhmann, 2010). Luhmann says that: "Love, as a medium is not itself a feeling, but rather a code of communication, according to the rules of which one can express, form and stimulate feelings, deny them, impute them to others, and be prepared to face up all the consequences which enacting such a communication may bring with it" (1986: n.p.).

Morgner in his study **The Theory of Love and The Theory of Society: The Remarks on the Oeuvre of Niklas Luhmann** shows that, in love there is an evolutionary change like Luhmann's says. In previous societies, love wasn't intimate or private thing but structural organization results in the change in the meaning and shape of love. Institutionalization of marriage and intimate love are two of them (Morgner, 2015: 398). Therefore, emotions don't take part only as biological of pyschological funtions but with social ones. It actually might show that living systems, pychic systems and social systems come together. Even if they are operationally closed because they are open to each other, interactional openness takes place. It can also be said that the relationship or with more appropriate concept **structural coupling** between the brain, the mind and the society come into existence. According to Luhmann, social systems that are structured by taking love as a base, includes communicative openness requirement (Luhmann, 2010). However, they aren't totally fixed and it might create problems. He explains that requirement with a basic example. Couples are like responsible to tell each other what they did every day. In modern marriages, the question "Why are you home so late?" cannot be answered like **that is none of your business** but Luhmann makes emphasis that, it is no guarantee that the question will be answered totally honestly (Luhmann, 2010).

When love becomes medium of communication in modern societies, the semantics of love change too. One of the examples can be the change in the meaning of sexuality to. In the previous stages while sexuality is related with procreation, in modern societies it doesn't. It becomes the meaning of commitment and this leads to the marriage and family. The choices in love then to be free to choose one's partner (Morgner, 2015: 399). However, the continuity of love is again based on marriage and divorce in that institutionalized system. Luhmann says that, "Love based on marriage is the result of that endeavour and the increased opportunities for divorce the corrective to it. In other words, it is left up to marriage whether love persist or not" (1986).

In general, love as a medium of communication through communication creates a kind of social operation. It is in Luhmann's terms a kind of self-observing operation. Love and institutionalization of love become a part of society because when communication develops from communication, self-referential autopoietic systems develop. In addition to this, to keep society alive communicating should continue.

REFERENCES

Baecker, D. (2006). Niklas Luhmann in the society of the computer. *Cybernetics & Human Knowing, 13*(2), 25-40.

Cankurtaran Öntaş, Ö., & Akçay, S. (2014). Niklas Luhmann'ın sistem teorisi ve teorinin sosyal hizmet disiplinindeki yansımaları. *Journal of Society & Social Work, 25*(2).

Çelik, C. (2007). Niklas Luhmann'da sosyal sistem olarak toplum ve modern toplumun karmaşıklığı sorunu. *Bilimname XII*, 2007/1, 51-74

Ertong, G. (2017). Niklas Luhmann'ın sosyal sistemler kuramı ve güven tartışmaları bağlamında sağlık sistemi. *Ankara Üniversitesi Sosyal Bilimler Dergisi, 2*(2).

Fuchs, S. (1999). Niklas Luhmann. *Sociological Theory, 17*(1), 117-119.

Görke, A., & Scholl, A. (2006). Niklas Luhmann's theory of social systems and journalism research. *Journalism Studies, 7*(4), 644-655.

Guy, J. S. (2018). Is Niklas Luhmann a relational sociologist?. In *The Palgrave handbook of relational sociology* (pp. 289-304). Palgrave Macmillan, Cham.

Hammond, D. (2010). *The science of synthesis: Exploring the social implications of general systems theory.* University Press of Colorado.

Lee, D. (2000). The society of society: The grand finale of Niklas Luhmann. *Sociological Theory, 18*(2), 320-330.

Luhmann, N. (1995). *Social systems.* California: Standford Publishing.

Luhmann, N (2010). *Love: A sketch.* Cambridge: Polity Press.

Luhmann, N. (1986). *Love as passion.* Cambridge: Polity Press.

Luhmann, N. (2006). System as difference. *Organization, 13*(1), 37-57.

Luhmann, N. (1995). *Social systems.* Stanford University Press.

Misgeld, D. (1994). Unlimited observation, unlimited observability: Niklas Luhmann's self-perpetuating systems theory. *Contemporary Sociology, 23*(1), 153-155.

Moeller, H. G. (2006). *Luhmann explained: From soul to systems.* Chicago: Open Court.

Morgner, C. (2014). The theory of love and the theory of society: Remarks on the oeuvre of Niklas Luhmann. *International Sociology, 29*(5), 396-404.

Parsons, T. (2013). *Social system*. England: Routledge.

Přibáň, J. (2010). Niklas Luhmann: Law, justice, society [Book Review]. *Modern Law Review, 73*(5), 893-897.

Probert, S. K. (2014). Introduction to systems theory [Book Review]. *International Journal of Systems and Society, 1*(1), 55-57.

Ritzer, G., & Stepnisky, J. (2014). *Modern sosyoloji kuramları* (Himmet Hülür, Trans.). Ankara: De Ki Basım Yayınları.

Seidl, D., & Becker, K. H. (2006). Organizations as distinction generating and processing systems: Niklas Luhmann's contribution to organization studies. *Organization, 13*(1), 9–35.

Seidl, D., & Becker, K. H. (Eds.). (2005). *Niklas Luhmann and organization studies*. Malmö: Liber.

Seidl, D., & Mormann, H. (2014). *Niklas Luhmann as organization theorist* (Vol. 125). Oxford: Oxford University Press.

Stichweh, R. (1999). Niklas Luhmann. *Klassiker der Soziologie, 2*, 206-229.

About Author(s)

ELIF OZUZ DAGDELEN is a research assistant in the Faculty of Science and Letters Department of Sociology at Başkent University. She graduated from Middle East Technical University Department of Sociology as the highest ranked student in 2016 and she did a minor in City and Regional Planning in the same university. She completed masters with thesis in Hacettepe University Sociology Department in 2018 and she continues her PhD education in Sociology in the same university. Since 2017, she has been working at Başkent University. Her fields of interest are digital sociology, sociology of media and sociology of communication.

7

SYSTEM APPROACH IN THE INTERNATIONAL RELATIONS

KIVILCIM ROMYA BILGIN

> On one hand an economy that grows so rapidly
> is intractably global. On the other hand, the
> current political system is intractably national.
> So there is a growing dichotomy between a
> global economy and locally based politics.
> — Walter Wriston

ABSTRACT

The formation of the system theory in the International Relations has been a tremendous stage in the field. After the system approach was introduced in the 1950s, it has been intensely used to analyze all kind of relations between actors and especially integration models of the states in the international system. As a result, the discussions on the systems concept and developing the system approach during the Cold War have dominated the field. Several political scientists have developed, adapted, and employed the systems concept for developing their own approaches and theoretical perspectives, even today. Thus, providing a general framework for the system approaches under the titles of the early period of system approach and system approach in structuralist theories will give a chance to show the place of the systems concept in the field.

SYSTEM APPROACH IN THE INTERNATIONAL RELATIONS

The formation of the systems theory in the field of International Relations has been a tremendous stage in the field. Before the concept of the system became widespread in International Relations, the concepts of the international community and world society were used extensively. The concept of system and the developing system approach, however, filled a theoretical gap in the literature. After the system approach was introduced in the 1950s, it has been intensely used to analyze the all kind of relations between actors and especially integration models of the states in the international system. As a result, the discussions on the systems concept and developing the system approach during the Cold War have dominated the study area in the International Relations. However, the social, political and economic changes experienced after the end of the Cold War have made other concepts popular such as governance or international society rather than the international system. Nevertheless, the concept of system and system approach in the International Relations has preserved its positions as a descriptive and analytical tool that can be used to overview the relations among states. It has been also a beneficial tool on both the studies of regional sub-systems which form parts of the international system and theory-building in international politics.

The system approach in International Relations has developed in light of improvements in other disciplines. The discussions on the system, especially in the discipline of Political Science, have inevitably influenced the discipline of International Relations. Several political scientists have developed, adapted, and employed the systems concept for developing their own approaches and theoretical perspectives, even today, political scientist as Karl Deutsch, David Easton, William Mitchell, Kenneth Waltz are called the founding fathers of the system approach in the Political Science and International Relations. Theoreticians contributing to the development of the system approach have provided the extensive discussion area and constructed the main axis of it, meanwhile, they have produced the different approaches especially on what is the function of a political or international system and how a system does structure at international area. Some have focused on the possibilities of analyzing the political or international environment as a system, while others have focused on the interactions between actors in a system. Discussing their contributions to the system approach is a compulsory condition for showing the place of that system idea in International Relations.

The most important feature of the systems concept is that even if a small part changes, the whole is affected and the change is inevitable. However, what these parts are and the relationship between them is expressed differently by different system theoreticians. As defined by Rapoport (1966, 1968), a system, "is a set of interrelated entities connected by behavior and history". For Fisher, Rapoport's approach on the system is quite extensive to include solar system or language system. Social systems, such as economics or politics, can also fit into the definition as well. Social perspective should be used to understand the political system and differentiate it from other systems. It is possible to define social systems as a class of individuals, families, institutions and their relations with each other by communication channels, influence, obligations (2010: 72). Hoffman gives more specific definition on system by focusing of its political side and defines it as "a pattern of relations among the basic units of world politics, characterized by the scope of the objectives pursued by those units and of the task performed among them as well as by the means used to achieve those goals and perform these tasks" (1961: 207). Beer and Ulam describe it as a "structure that performs a certain function for a society" and "produces certain outputs for the society such as legitimate policy decisions" (1968: 21-22). As can be seen in the definitions of system, the relationship between the systems concept and other concepts such as entity, unit, output or objective reveals the basic nature, structure, and functioning of a system. All definitions give a clue that political systems are dynamic and complex processes rather than repeated static patterns.

Political scientist such as David Easton, Karl W. Deutsch, Gabriel Almond, and Morton Kaplan are important figures in influencing the conceptualization of political systems in the field of International Relations by taking political systems in broader theoretical terms. These names, mainly influenced by behaviouralism, are referred to as early representatives of the system approach in International Relations. However, Kenneth Waltz, Robert O. Keohane, and

Immanuel M. Wallerstein developed a structuralist perspective by referring the phenomenon of system in International Relations and became the representatives of the system approach in the field. Each of them made different readings about the system in the field from realism, liberalism, and economic-political perspective. Thus, the system approach in International Relations has developed under different theoretical approaches but all named as structuralist theories. Providing a general framework for the system approaches under the titles of the early period of system approach and system approach in structuralist theories will give a chance to show the place of the systems concept in the field.

Early Period of System Approach

To comprehend how the system approach has evolved in the field, at the first hand it should be used a historical perspective. For that reason, the prerequisite to understand as to how the discussion on behaviouralism has emerged and developed. Behavioralism has roots intellectually and theoretically in behaviorism, a psychological school founded by James B. Watson and inspired by Jacques Loeb's study (Lasswell, 1950: 553). Following the studies of John B. Watson, the founder of behaviorism, B. Skinner took behaviorism to a different level in Psychology and soon behaviorism became effective in other disciplines. As a result, various opinions have emerged in the literature regarding the definition of behaviorism that manifests itself in different disciplines such as Political Science and International Relations. Discussions on behaviorism are especially crucial because it sets the foundation for behaviouralism. According to Skinner behaviorism "is not the science of human behavior: it is the philosophy of that science" (1976: 3). What Skinner underlined that behaviorism should be treated as a scientific understanding is the main effect of behaviorism to behaviouralism.

American political scientists after the Second World War have designed behaviouralism as a scientific way of understanding to the analyze and explain political phenomena. In this way, Behaviouralism came to be something bigger than the study of political behavior, yet political behavior was always its main focus. Behaviouralism has turned out to be a set of orientations, procedures, and methods of analysis by not remaining confined to the study of individual-based political behavior (Wogu, 2013). The debate about what behaviouralism is and its definition continue even today, despite that fact it is possible to reveal the basic structure that constitutes the character of behaviouralism. The basic understanding of behaviouralism in social science is that it can be scientific in terms of methods and explanation as in natural sciences. Behaviouralism insists that observable behavior, whether on an individual level or a social level must be the focus of any study. According to behaviouralism, scientific models can be developed for the dynamic systems of human beings and their interactions with their environment. This explains why the systems concept and systems approach have become predominant in social science such as Political Science and International Relations.

Behaviouralism has been immediately popular and effective in the discipline of International Relations in the 1950s. **A Study of War** written by Quincy Wright (1942) recorded as earliest study in the field was conducted under the effect of behaviouralism. However, after World War II, behaviouralism has became dominant and one of the founding debates of the field has emerged. The realism-idealism debate in the previous period has been defined as traditionalism and then a new debate between behaviouralism and traditionalism has launched in the field. While realists-idealists have argued over the nature of international politics, the focus of the debate between behaviorists and traditionalists is about to construct the appropriate methodology in the field. The posture of behaviouralism and its methodological perspective in the field was quite different from other disciplines, because it has been referred to as the way to study politics or other social phenomena that focus on the actions and interactions among units by using scientific methods of observation. While traditionalists have emphasized the history, law, philosophy, and other traditional, non-quantitative methodologies to comprehend all actors in international politics, behaviorists have favored hypothesis, empirical testing, and model building in the study of political processes (Viotti & Kauppi, 2012: 442-443). Even today, the debate between parts of the field continues and allows different approaches to be raised in current issues in the field.

David Easton has been one of the early political scientists being introduced to the behaviouralism and system approach to politics. Easton's influential study named as **The Political System: An**

Inquiry into the State of Political Science was published in 1953. The study had a profound impact on Political Science and International Relations. Easton's famous definition of politics is "the study of the authoritative allocation of values for a society" (1953: 129). His politics definition was a part of a general approach of him on institutions and processes in the political system. Easton has preferred to select the political system as "the basic unit of analysis" and defined a summarily political system as "a set of interactions" (1953: 21). After that Easton defined a system as "any set of variables regardless of the degree, of inter-relationship among them" (1965: 21). However, his definition and approach on politics and system has raised the question of how the methodology should be at the first stage. In that perspective, Easton has developed his perspective and drawn the basic principles of behaviouralism in Political Science. Easton's definition of political behaviouralism as a science of politics has been based upon natural sciences' methodological assumptions. In this context, the eight basic principles as given by Easton are as follows (1962: 7-8):

1. Regularities: In political behavior, there are discoverable uniformities. Thanks to them, political behavior might be described in generalizations or theories with explanatory power.
2. Verification: The methods used in natural sciences can be adapted to political science, and thus, hypotheses of political behavior can be empirically tested.
3. Techniques: The means necessary to acquire and interpret data, as in the natural sciences, should be examined and validated.
4. Quantification: Behavioral perspective requires collection and interpretations of data which is not obtainable under the traditional perspective.
5. Values: Scholars should not confuse empirical reality with their values. Because ethical evaluation and empirical explanation are totally from each other.
6. Systematization: There is a close relationship between research and theory. Without the conceptual framework of the data masses, the theory would be sterile without data.
7. Pure science: The comprehending political behavior helps to use political knowledge to solve practical problems in society.

8. Integration: Including political science, all social sciences deal with all human condition. Interndisciplinary orientation for political research may be utilized.

Easton's principles were the basis of Eason's political system approach. With adhering to his own principles, Easton produced an improved political system model as stated in Figure 1. Easton focused on the interaction in the system and the exchange of the system and its environment. From Easton's point of view, a political system should be distinguished from other social systems. According to him, the best way to do this is to perceive a political system as a self-contained entity that has characteristics that distinguish it from the environment. To this end, political systems' fundamental units and also its boundaries should be described (1957: 384-386).

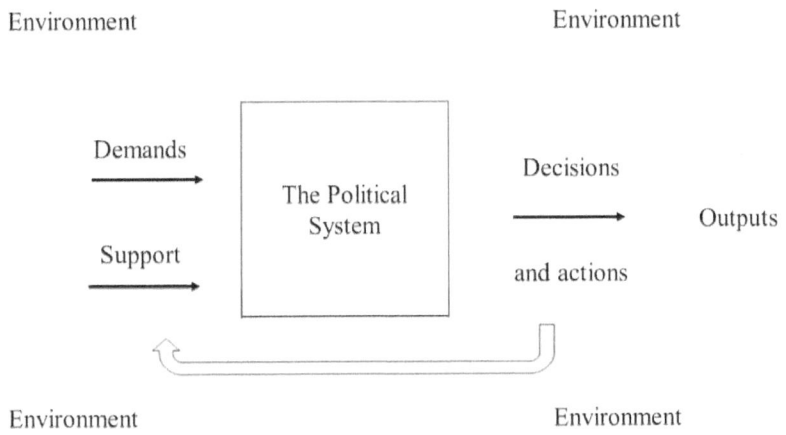

Figure 1: Easton's Political System Model
Source: Easton (1957: 384)

Easton defined five types of action which are the elements of all political systems. These are legislation, administration, adjudication, the development of demands, and the development of support and solidarity. All these were defined and specified as input and output requirements for any political system (1957: 384-386). The system can not work without inputs; the work done by the system can not be identified without outputs. Inputs are the demands and the support of the members of the system, on the other hand, outputs are the decision and action of the authority (1957: 385-386). Easton gives special attention feedback which is communicative process producing action in the political system. Feedback in political system vitalizes the system connecting outputs (authoritative decisions) to inputs (support and demands). Feedback also gives a chance to a political system to regulate itself for surviving (Easton 1965: 116). This model, which seems simple at first sight, is actually important in terms of being the first comprehensive framing of the political system.

Easton is not the only political scientist who has dealt with the system. Like Easton, Gabriel Almond (1956) desired to develop a political system approach, but Almond aimed at enhancing a more universal and specific system approach to include developing countries. Almond suggests that all political system exist that in the both domestic and international environment. Almond also stresses the interdependence between political and social systems which is the base of a comparative perspective. In 1960, Almond wrote a book with James Coleman named as **The Politics of The Developing Areas.** They have defined political systems as "that system of interactions to be found in all independent societies which perform the functions of integration and adaptation" (Almond & Coleman, 1960: 7). Then, Almond and Coleman were firstly compared and analyzed the political systems of developing countries by using a common set of categories. They have benefited from sociological and anthropological theories in order to find proper categories for comparing the developing countries. Indeed, Almond and Coleman have actually adapted an old language to a new situation. For example, instead of using the concept of state, they used the political system, because the concept of state is limited by its legal and institutional meanings. Instead of using powers, because of its legal connotations, they preferred to emphasize functions; and also, instead of using the concept of institutions, that directs thinking toward formal norms, they prefered to use the concept of structures (Fisher, 2010: 76).

All political systems have four common characteristics, according to Almond and Coleman, and in terms of which political systems may be compared. First, all political systems, even those simplest, have political structures. All these structures have designed for certain functions to perform. Second, in all political systems, the same functions are being performed. Third, all political structures have an ability to be multifunctional, whether they are in primitive or modem societies.

Finally, in the cultural meaning, all political systems are mixed as well (1960: 11). These are just characteristics of any political system. On the other hand, in all political system, there are certain structures (such as interest groups, legislatures, executives, bureaucracies and courts) and in every political system, there may be differences between structures. Even though being differences in the systems and structures among different countries, but all systems perform almost the same political functions (1960: 18). Almond called this approach the functionalist approach for comparative political science and then in 1966, he further developed his approach by writing a book with B. Powell titled as **Comparative and Politics: A Developmental Approach**. Almond integrated the structural functionalism's theoretical framework with political culture, political socialisation, and political development. The main issue is that Almond's discussion of functionalism through system theory has fostered structuralist and functionalist debates in international relations. According to Hix and Høyland, through Easton and Almond's studies, in the mid-1950s and 1960s four characteristics of the political system have developed in the field as follows (2011: 12–13):

1. There are clearly defined institutions for collective decision making and also there are rules for governing the relations among them.
2. Through the political system, citizens try to find ways for realizing their political aspirations, either directly or through proxy entities such as interest groups and political parties.
3. The collective decision mechanism in the political system has a strong influence on the allocation of economic resources and values.
4. There is constant contact between political outcomes, new demands from system, new decisions, and so on.

Easton, Almond, and others started a major shift in the 1950s and 1960s with their opinions regarding the system in political science but they were not the only ones. Other influential studies were published in 1957. Morton A. Kaplan has published his most known study which is **System and Process in International Politics**, and then Karl Deutsch et al. have published **Political Community and the North Atlantic Area**. These studies have fostered the behaviouralism trend in International Relations and after their publishing, the use of systems concept and system approach has extremely become more popular. Kaplan's study with a behavioralist attitude has enabled the system approach to gain a solid place in international relations. He preferred to name for system as "system of action" which described as "a set of variables so related" (2005: 20) and then Kaplan defined the international system as "action taking place between international actors" (2005: 33). Kaplan's system approach provided an opportunity for new discussions in the field and encouraged new methodological discussions. To analyze the international system, according to Kaplan, the state of an international system or its subsystems should be described by considering particular variables: the essential rules of the system, the transformation rules, the actor classificatory variable, the capability variables and the information variables (2005: 24). In his opinion, each of these variables allows an understanding of the working principle of the international system and represents a very important part of the system. These variables can be summarized as follows (2005: 24-26):

1. The transformation rules are the laws of change of the dynamic system. They specify linkages under given parameters of condition so that the future of states of the system becomes predictable.
2. The actor classificatory variable is about to specify the structural characteristics of actors which may change for each actor. For example, a nation state's characteristic as an authoritarian or democratic would have a various effect on her behavior.
3. The capability variables points out the physical capability of an actor to carry out actions in the system. Even though it does not directly indicate a general power to act, it is quite related to various determinative factors such as territory, population, industrial capacity, military force, etc.
4. The information variables are about an actor being aware of his capacity to do or not to do something.

After Kaplan sets out these variables or the functioning of the international system, he seeks to answer the question of how the international system is structured. Kaplan's international system consisted of states interacting in six possible models (2005):

1. The Balance of power is a system that does not allow any of the nation states with a minimum number of five and approximately equal powers to prevail. There are a few essential rules for actors to protect balance of power system:
   - Increase capability but negotiate rather than fight.
   - Fight instead of failing to improve capability.
   - Stop fighting instead of eliminating an essential actor.
   - Oppose any alliance or single actor who tends to take a leadig position within system
   - Constrain actors who abide by supranational organizational principles
   - Enable defeated or limited essential national actors to re-enter the system as suitable role partners or act to bring some previously unessential actor into the classification of the essential actors.

2. The loose bipolar system is effective when two powerful nations organize the other nations into their two respective competing blocs or groups.

3. The tight bipolar system is a bi-polar system in which the two major powers are leading their respective ally powers blocs. The Loose Bi-polar System could easily become a tight bi-polar system.

4. The universal system is a conceptual model that organizes the nations into a federal system.

5. The hierarchical system will come into being when a single powerful super may bring. In this system super power control other nations.

6. The unit veto comes into existence in which each state is just as strong. In this system, each nation or group can destroy all others, there is mutually assured destruction.

Kaplan has established a strong system approach in the field of International Relations with six models. However, Kaplan's system approach has received serious criticism. For example, Robert J. Lieber (1972) found Kaplan's system approach limited and inadequate, especially because he did not pay attention to political institutions and important domestic variables. Kenneth Waltz (2010) criticized Kaplan's too much emphasis on state and for that reason Waltz described Kaplan's system approach as a reductionist. Even though all critics,

Kaplan's system approach has initiated a deep discussion about the roles and relationships of the actors in the system and took its place as a comprehensive analytical tool.

Karl W. Deutsch further deepened system approach with his views in 1963 with his book published, **The Nerves of Government: Models of Political Communication and Control**. Deutsch has used cybernetics model to explain the flow in the international system. Deutsch states that the viewpoint of cybernetics is that every organization like governments is held together by communication in where information-processes are mainly represented. Communication is the ability to transmit messages and react to them (1963: 76-77). In Deutsch idea, cybernetics is about communication process in systems or organizations. Deutsch has made an important contribution in terms of bringing the concept of communication which is ignored in international relations into the field.

From Easton to Deutsch, the early system approaches considered a system "as a totality composed of its parts". The international system was consisted of nation states and only their 'interactions' were at the core of systems studies (Tayfur, 2000: 13). Therefore, early system approaches were easily understandable even if they were not methodologically and ontologically in totally sufficient. However, they have seriously failed to address different critical variables such as globalization, transnational actors and economic crises. Their shortcomings have then completed those who address the systems concept with more specific theoretical perspectives in the field.

System Approach in Structuralist Theories

Each system theoreticians uses an indefinitely different definition for the systems concept, but all agree that the system as an analytic concept requires a study on the interaction of states within a particular structure. On the other hand, they did not have a clear attitude to approach the systems concept from more strict theoretical perspective. They were the structuralists who placed the system strongly in the field epistemologically and ontologically, although they did not concern placing the system understanding directly in the field. Kenneth Waltz is a benchmark author who successfully integrated realist theory with the system (Telo, 2009: 48). In spite of the strength of early approaches and the continuing theoretical impacts, criticisms of the early system approach have allowed new approaches to the system in international relations. For the first time, Kenneth Waltz, in his book **Man, The State and War** published in 1959, deals with war as a problem of analysis level by which he builts a useful structure for handling the international system. Waltz never uses the term **levels of analysis** anywhere in his book. Waltz has preferred to use the concept of **images** of international relations to refer the international system and structure of it. Conceptualization of level of analysis has been given his approach by the others and he developed with his later studies.

In his study, Waltz develops the approach of understanding wars through analysis of the system by images, based on the idea that one must understand the system to understand the major causes of wars. From that point of view, Waltz suggests that there are three images or in other words levels of analysis that can be utilized in the study of how wars occur. In the first level, the assumption is that the egotistical human nature causes wars. This level of analysis suggests that we do not need to go further than the personal attributes of policymakers to appreciate the causes of wars (1959: 16). In the second level of analysis, there is a shift from an individual level to a more collective level. In this level, the state and the internal constitution of the state, such as its ideological underpinnings or political regime, have been mattered. Waltz argues that any explanation presented at these first two levels is not entirely sufficient (1959: 81-82). Therefore, it is necessary to specifically look at the third level which describes the framework of world politics as anarchic structure (1959: 160)

By the time Waltz's system approach has been differentiated significantly from both Kaplan's and Deutsch's approaches. In 1979, Waltz further developed his views with the book **Theory of International Politics** to put an international system theory and focused on the anarchic structure of the international system. In Waltz view, "a system is composed of a structure and interacting units. The structure is the system-wide component that makes it possible to think of the system as a whole" (1979: 66). There are several characteristics of that system which make it complex. Waltz argues that states are the only important actors in the international system and they behave according to their place in the system. He recognizes other actors such as international organizations or non-governmental organizations but says that they do not have a powerful effect on the system like powerful states. Systems do not act as a whole, but agents and agencies act constantly. But the actions of agents and agencies are influenced by structure of system. Structure influences actions of agents and agencies within the system but does so indirectly. The effects are created in two ways: through the actors' socialization and competition among actors. Those two mechanisms exist in communities in international politics to be the same. The structure as concept is based on the idea that units' behavior combined differently and produce different results in their interaction (1979: 74-81). After explaining relationship between the actors, Waltz reveals the three basic features of a political structure which means the form of the relationships of the states composing international system (1979: 89-98):

1. Ordering the System: The international systems are anarchic and decentralized. Absence of system-wide authority and administrative power leads to disorganization and lack of central organization.
2. The Character of Units: The international system consists of sovereign states that are similar functions. States have the same functions but distinctions among them arise principally from their varied capabilities.
3. The Distribution of Capabilities: The most important feature of the system is the distribution of capabilities. While capability is about units, the distribution of capabilities is a system-wide concept. Historically, there has been no international system other than bipolar and multipolar in terms of the distribution of capabilities.

Waltz argues that bipolar systems represent the most ideal international political system exemplified in history, and speaks of the advantages of bipolarity, especially to the great powers. For example, in these systems, interdependences among states are greatly diminished, and the great powers are based on their potential rather than their allies. So, they can follow independently-determined strategies. Possible dangers and responses can be easily calculated by the competing party (1979: 171-172). At the system-level model, the distribution of capabilities has a crucial role in international politics. Therefore, there are two steps to analyze the effects of the structure on the foreign policy choices of states: First, an examination of each state's place in the system, second, an examination of the nature of the system.

The most important contribution of Waltz was the employment of system approach in the field of international relations and hence the revision of some of the basic concepts in traditional realism. On the other hand, the 1970s is a period in which important studies have been made for the development of the system approach from a liberal perspective. Another study conducted during this period belongs to Robert O. Keohane and Joseph Nye. Even though they did not specifically claim to bring a new perspective for the system approach, their studies referred to how the international system works and structures. In 1977, they published their influential study named as **Power and Interdependence: World Politics in Transition**. Along with the 1984 publication of **After Hegemony: Cooperation and Discord in the World Political Economy** by Keohane both studies are among the leading texts in neoliberalism, respectively.

Keohane and Nye have suggested using the concept of complex interdependence to explain the sophisticated structure of global politics while accepting some of the basic assumptions of realism. According to them, the concept of interdependence means mutual dependence and in world, politics refers to conditions marked by mutual effects of different countries or actors. These effects are often derived from international transactions such as flows of people, commodities, finance and communication across international boundaries. In this system of **interdependence**, states and other transnational actors cooperate because of their common interest. There are few direct results of this cooperation like prosperity and stability in the international system. On the other hand, the relationship between the actors including states as well as other transnational actors in this system is characterized by competition as well (2012: 6-8). By defining the concept of interdependence, Keohane and Nye characterize three characteristics of **complex interdependence**. The first one is that there are multiple and layered channels connecting societies. The second one is that there is an absence of hierarchy among issues in the interstate agenda. The first two characteristics are directly related to the third characteristic of interdependence which is diminishing the importance of military force (2012: 20-24). Complex Interdependence stresses the complex ways in constructing relations among states and other actors. This is the result of increasing ties and ultimately the transnational actors become mutually interdependent, vulnerable to each other's actions and sensitive to each other's needs.

Keohane in his seminal work, **After Hegemony**, provides a more compelling justification for the role of international institutions in world politics. He claimed that even the absence of hegemonic power, cooperation might be possible between states because international institutions can alleviate collective action. They can create feasible conditions for cooperation by monitoring state behavior or by linking multiple issues together (Keohane, 1984: 88-98). Like Waltz, Keohane uses the systemic level of analysis in examining the international system specifically on the relations among advanced-market economy countries. The approach of Keohane and Nye on the actors' interactions in the international system through interdependence, and consequently Keohane's emphasis on the diminishing role of hegemony in the international system is a debate on the systems concept from a neoliberal perspective. Keohane being aware of the importance of understanding the system argues that the Systemic theory is crucial because, before understanding action itself, the meaning of action must be understood (1986: 193). Waltz and Keohane's approaches to the functioning of the international system are the focus of the debate between neo-realism and neo-liberalism. For this reason, the debate on the international system in the field should not be overlooked in these two names.

Immanuel M. Wallerstein expressed modern world-system analysis in his seminal paper **The Rise and Future Demise of the World Capitalist System: Concepts for Comparative Analysis** in 1974 and his outstanding book **Modern World System: I: Capitalist Agriculture and the Origins of the European World-Economy in the Sixteenth Century** in 1976. The analysis of Wallerstein is quite different ontologically and methodologically from Waltz's and Keohane's approaches because of its neo-Marxist theoretical understanding on which it is based. The world system analysis offers a wide theoretical perspective to comprehend the historical evolutions and changes in the emergence of the modern capitalist world-system. The analysis is also an attempt to understand the existence of the current global system. That's why Wallerstein needed a unit of analysis huge enough to embody all causes of structural social changes (Terlouw, 2018: 85). In this context, unlike previous theoreticians who contribute the World system approach, Wallerstein tries to develop an understanding and an analysis model that includes the concepts of culture and ideology with a neo-Marxist perspective that bases economic conditions. Indeed, Wallerstein's analysis of the world system is a critique and an opposition to capitalist modernization. In doing so, Wallerstein develops an economic and historical understanding and draws clear boundaries.

According to Wallerstein, three types of social systems have existed so far. First, are mini-systems "an entity that has within it a complete division of labor and a single cultural framework". These are very small with a single culture and polity. Only some isolated indigenous tribes can be classified as mini-systems (1974: 390; 1979: 155). The other systems are a World-empires and world-economies. The main difference between them is about their political structure. World-empire is characterized by central structure and there is a single political system over most of the area in the world. On the other hand, in world-economy, there might be various political structures, but it is dominated by one economical structure (1976: 229-230). Wallerstein specifically focused on a world systems concept. A world-system that is named as a world economy by Wallerstein means an integrated market structure rather than a political center. In this system, two or more regions are interdependent on food, fuel, and protection, and there is a competition between two or more polities for domination over the system (Goldfrank, 2000). He depends on the world-economies as a system and defines the world system in the final. For him, a world system "is a social system, one that has boundaries, structures, member groups, rules of legitimation, and coherence. Its life is made up of the conflicting forces which hold it together by tension and tear it apart as each group seeks eternally to remold it to its advantage. It has the characteristics of an organism, in that is has a life-span over which its characteristics change in some respects and remain stable in others." (1974: 347). Wallerstein constructs the world system on an economic basis but also emphasizes its ideology and cultural infrastructure. Thus, he sets up a political, sociological and historical analysis of the world system as a whole. Because the economic, political and cultural events and their relations, which are perceived as independent from each other, are interconnected in a large structure and the relationship between them is systemic. Therefore, the world system is composed of a single and large society whose functional parts are interconnected by various fields such as economy, politics, and culture. In this context, modern world system analysis historically investigates the structural relations among different societies during the same historical terms. As a result of this historical analysis, the world system analysis, since the 16th century, not only functional but also occupational – but geographical – is characterized by the international division of labor has argued that a world system (1974: 348). Wallerstein emphasizes that a world system is a "multicultural territorial division of labor in which the production and exchange of basic goods and raw materials are necessary for the everyday life of its inhabitants" (1974: 347).

The foundation of the new capitalist world system is based on an international division of labor which determines relationships between different regions as well as the types of labor conditions within each different region. According to this model, there is a direct relationship between the type of political system and the place of each region in the world economy. In this context Wallerstein proposes four different categories in order to describe each state or region's position within the world economy. These are the system core, periphery, semi-periphery and external area. The concepts of core and periphery are not Wallerstein's concepts and are used by the dependency school,[8] Wallerstein develops these concepts in order to analyze the emergence of capitalism in Europe. The core areas historically have engaged in the most advanced economic activities such as banking and manufacturing. The periphery has provided raw materials to the core's economy. There are also semi-peripheral areas which are in between the core and the periphery. The semi-periphery areas are involved in a mix of production activities. While some of them are associated with the core, some of them are associated with periphery (Viotti & Kauppi, 2014: 201). Some of the semi-peripheral areas were core-areas of earlier versions of a given world economy. Some were peripheral areas which were promoted later on. In other words, semi-peripheral areas are in some way a buffer zone between core and periphery. That's why, the semi-periphery is the most dynamic part of the World-system (Terlouw, 2018: 90) External areas are very limited and they maintained their economic systems and mostly managed to remain outside of the modern world economy. Despite economic

---

8   The Latin American-based dependency school sees the spread of capitalism as responsible for underdevelopment. According to the school of dependency, capitalism does not contribute to the development of less developed countries, but their underdevelopment.

and political power inequality among all categories, there is constant relation among them. World system analysis emphasizes the social structure of global inequality. It is also a powerful tool for analyzing how the worldwide inequalities in our world formed and perpetuated (Terlouw, 2018: 94). Besides by using this analysis, it can be described the effect of economic flows on the separation of states.

Waltz's neo-realist, Keohane's neo-liberal and Wallerstein's neo-Marxist theoretical perspectives, although they draw different frameworks about how the international system works, are not only contributed the development of the system approach but also as a whole in international relations.

Conclusion

The systems concept is very comprehensive and it is a powerful analytical tool which makes it easy to find a place in different fields. The systems concept has gained its place in political science and then in international relations with the effect of behaviouralism. By the time, this place has developed on the various theoretical grounds and has become an understanding and a way of thinking by going beyond a concept. This understanding and the way of thinking about the system has made itself felt especially in the debates carried out by structuralist approaches.

The argument is at the center in many systems approaches that the structure of the system should be defined in terms of the functional differentiation among states as homogeneous or heterogenous, ordering principle as anarchy or hierarchy and distribution of power among states (Mansfield, 1994: 7). Thanks to the system studies, scholars in the field have started mainly to consider the interaction among the states as well as examining the actions of individual states in international relations (Tayfur, 2000: 13). Besides that, there are other several reasons to benefit the system in International Relations. First, system gives a chance to identify the arrangements among international actors in which interactions among them are complex. Second, system is useful analytical tool to define a particular and to explain the behaviour of particular states in this particular order even though theoretical explanations based on the system approach may be insufficient to explain all the actors at any time, every situation and interaction at the macro level. Although the theoretical boundaries of the system approach in International Relations seem to have become clear over time, it is possible that the dynamic structure of international system can bring about new theoretical discussions. On the other hand, it is obvious that this dynamic structure will also provide the opportunity to be an analytical tool which is always used in the micro level studies and discussions about the system approach.

REFERENCES

Almond, G. (1956). Comparative political systems. *The Journal of Politics, 18*(3), 391-409.

Almond, G. A., & Coleman, J. S. (1960). *The politics of the developing areas.* New Jersey: Princeton University Press.

Baum W. M. (2005). *Understanding behaviorism: Behavior, culture, and evolution.* 2nd edition. Oxford: Blackwell Publishing.

Beer, S. H., & Ulam A. B. (1969). *Patterns of government.* New York: Random House.

Deutsch, K. W. (1963). *The nerves of government: Models of political communication and control.* New York: The Free Press:

Easton, D. (1953). *The political system: An inquiry into the state of political science.* New York: Alfred A. Knopf.

Easton, D. (1957). An approach to the analysis of political systems. *World Politics, 9*(3), 383-400.

Easton, D. (1962). Introduction: The current meaning of 'behavioralism' in political science. In J.S. Charlesworth (Ed.), *The limits of behavioralism in political science* (pp.1-25). Philadelphia: American Academy of Political and Social Science.

Easton, D. (1965). *A systems analysis of political life.* New York: Wiley.

Fisher, J. R. (2010). Systems the theory and structural functionalism. In J. T. Ishiyama, & M. Breuning (Eds.), *21 st century political science* (pp. 71-80). Texas: Sage.

Goldfrank, W. L. (2000). Paradigm regained? The rules of Wallerstein's world system method. *Journal of World-Systems Research, 6*(2), 150-195.

Goodman, J. S. (1965). The concept of system in international relations theory. *Background, 8*(4), 257-268.

Hoffmann, S. (1961). International systems and international law world politics. *The International System: Theoretical Essays, 14*(1), 205-237.

Kaplan, M. A. (2005). *System and process in international politics.* Essex: ECPR Press.

Keohane, R. O. (1984). *After hegemony: Cooperation and discord in the world political economy.* New Jersey: Princeton University Press.

Keohane, R. O. (1989). Theory of world politics: Structural realism and beyond. In *International institutions and state power: Essays in international relations theory* (pp. 35-73). Boulder, CO: Westview Press.

Keohane, R. O., & Nye, J. (2012). *Power and interdependence: World politics in transition.* Fourth Edition. Boston: Peason.

Lasswell, H. (1950). Psycholgy and political science in the USA. In *Contemporary political science: A survey of methods, research and teaching* (pp. 526–537), No 427. Paris: UNESCO.

Lieber, R. J. (1972). *Theory and world politics.* Boston: Little Brown.

Mansfield, E. D. (1994). *Power, trade and war.* New Jersey: Princeton University Press.

Rapoport, A. (1966). Some system approaches to political theory. In D. Easton (Ed.), *Varieties of political theory* (pp. 129-142). Englewood Cliffs, NJ: Prentice Hall.

Rapoport, A. (1968). General systems theory. In D. L. Sills (Ed.), *International encyclopedia of the social sciences* (pp. 452-457). New York: Macmillan.

Sjöstedt G. (2004). The systems approach in research on international relations: The WTO negotiations. In M. O. Olsson, & G. Sjöstedt (Eds.), *Systems approaches and their application* (pp. 253-265). Dordrecht: Springer.

Skinner, B. F. (1974). *About behaviorism.* New York: Vintage Books Edition.

Tayfur, M. F. (2000). Systemic-structural approaches, world-system analysis and the study of foreign policy. *METU Studies in Development, 27*(3-4), 265-299.

Telò, M. (2016). *International relations: A European perspective.* New York: Routledge.

Terlouw, K. (2018). World system analysis. In R. C. Kloosterman, V. Mamadouh, & Pieter Terhorst (Eds.), *Handbook on the geographies of globalization* (pp. 84-95). Massachusetts: Edward Elgar Publishing, Inc.

Viotti, P. R., & Kauppi, M. V. (2012). *International relations theory.* Boston: Pearson.

Wallerstein, I. (1974). The rise and future demise of the world capitalist system: Concepts for comparative analysis. *Comparative Studies in Society and History, 16*(4), 387-415.

Wallerstein, I. (1976). *The modern world system: Capitalist agriculture and the origins of the European economy in the sixteenth century.* New York: Academic Press.

Waltz, K. N. (1959). *Man, the state and war.* New York: Columbia University Press.

Waltz, K. N. (1979). *Theory of international politics.* Illinois: Waveland Press Inc.

Wogu, I. A. P. (2013). Behaviouralism as an approach to contemporary political analysis: An appraisal. *International Journal of Education and Research, 1*(12), 1-12.

Wolfram, F. H. (1965). The international system: Bipolar or multibloc?. *The Journal of Conflict Resolution, 9*(3), 299-308.

Wright, Q. (1942). *A study war.* Chicago, IL, US: University of Chicago Press.

About Author(s)

KIVILCIM ROMYA BILGIN is an Assistant Professor in the Faculty of Communication at Baskent University. She completed her undergraduate education at Başkent University, Department of Political Science and International Relations. In 2008, she graduated from Middle East Technical University, Department of Political Science and Public Administration, and in 2015, she received her PhD degree from Turkish Military Academy on International Security and Terrorism. Since 2016, she has been working at Başkent University. Her interest includes studies on war and peace and the relationship between strategy and media.

8

Artificial Intelligence And Intelligent Systems

Ali Serhan Koyuncugil
And Nermin Yenikose

> *"I visualize a time when we will be to*
> *robots what dogs are to humans, and*
> *I'm rooting for the machines."*
> *— Claude Shannon*

Abstract

In this chapter, artificial intelligence (AI) and intelligent systems (IS) are discussed in details. AI methods, AI and IS examples are given with their methodological roots. Furthermore, vital elements of statistical data analysis are given and their extensions on artificial intelligence, intelligent systems, machine learning, deep learning, big data analytics and data mining are explained in details with method examples.

Introduction

While computational accelaration is getting day by day, data from different sources in different formats increasing exponentially. Therefore, the need for understanding, processing, modeling and valuating this enormous data has been caused new approachs, disciplines and techniques such as Machine Learning, Deep Learning, Artificial Intelligence, Big Data Analytics, Data Mining and more. Actually, it is hard to follow this sequential methods flood, because in every new day almost the same things are representing us with new labels. On the other hand, all these **brand new data technologies** have the same roots or based on classical theories. In this study, artificial intelligence and intelligent systems introduced from decision theory point of view which is the base of artificial intelligence.

Digitalization

One of the key factors for AI is digitalization. Because, the main source for Artificial Intelligence and Intelligent System development is **data**. Digitalization is the key factor which moves AI forward. Therefore, digitalization chronology in Figure 1 is given below with important developments (Fayyad, Piatetsky-Shapiro, & Symth, 1996; Koyuncugil, 2013).

Rapid improvement of Information Technologies turns everyting IT dependent. When we talk about this transformation, we are talking about from 1950's until now. General chronology of digitalization is given below from IT development point of view.

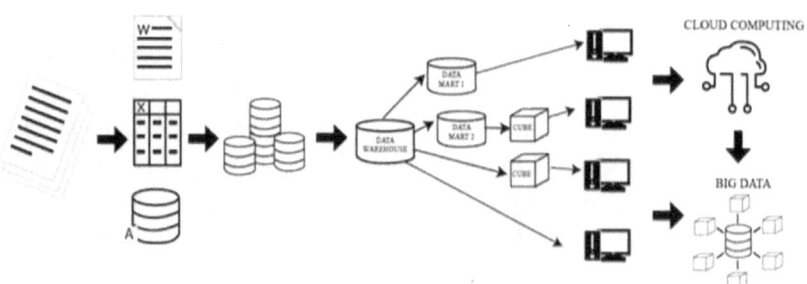

Figure 1: Chronology of Digitalization

1950's

The first computers were only used in censuses.

1960's

Data collections and

Database creation

1970's

Relational data model and

Relational database implemantations

1980's

Relational database systems grow up and

Implemantations of database management systems (spatial, scientific, engineering)

1990's

Records have been kept in papers until early 1990's. In 1990's office programs have been started to use the records instead of papers. Office programs made easier to store the documents than the papers. On the other hand, office programs came up with another problem: organization of data. Then, the solution came up with another IT solution as databases in mid 1990's. Therefore, firms began to keep their records in databases instead of separated PC's and of separated files. Databases increased the accesability of the records or data. Consequently, it is started to query that how big data collecting from daily life can be valuating.

2000's

Data is collecting from daily life stored in data warehouses. Data warehouses provided data source which is ready to analysis. Furthermore, not only structure but unstructured data started to use in data warehouses as well. Datawarehouses became a base for data mining. Furthermore, datawarehouses triggered another idea which is big data analytics.

2010's

In 2010's Cloud Computing began to use and it makes unstructured data use widely via put software, hardware and data together in the same eco-system. So, all kind of data SQL type and No-SQL type began to use effectively. Then, database concept matched with another concept File System. Database and File System match made big data a vital part of our daily lives. Then, the big data allowed to develop more accurate AI models with low error predictional models.

FROM HUMAN DECISION MAKING PROCESS TO ARTIFICIAL INTELLIGENCE

Decision making process simply based on three sequential steps as **data**, **information** and **decision** is given in Figure 2. However, sometimes and additional step as knowledge adds the process. On the other hand, knowledge covered by decision (Koyuncugil, 2013).

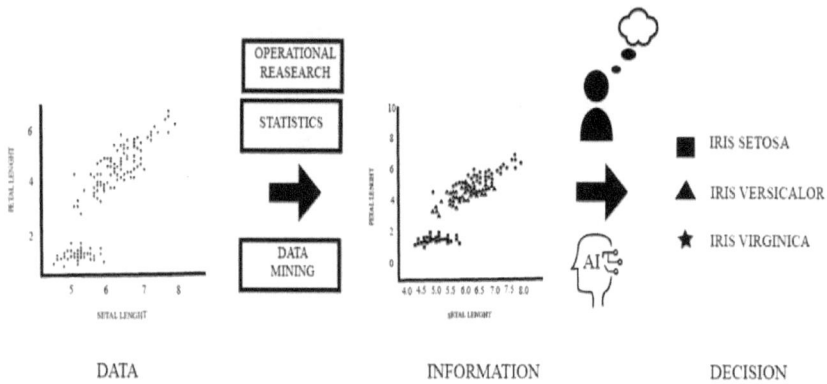

Figure 2: Decision Making Process

Data is the rawest situation of the information such as observation or administrative records. Mostly, data needs to pre-process such as cleaning, transformation before processed to be information. Cleaned (and transformed) data processed with decision making methods as operational research, statistical data analysis and in last decades with data mining. The main important part related with artificial intelligence is between information and decision steps. In traditional decision making process, only human mind evaluated the information and then turn it into decision. But Artificial Intelligence has been changed this tradition. Mainly, AI uses a similar evaluation process of human mind to make a decision from information:

1. Experience for repeated situations,
2. Predictions for new situations.

Both situations mean processing of past data via predictive model. So, simply AI uses the steps given below like human thinking:

1. Learning from data (training),
2. Verification (testing),
3. Developing a decision (prediction) model, and
4. Use new inputs (data) for developed model.

Some Vital Statistical Elements

In this section, some vital statistical elements are introduced which are linked to intelligent systems. Because statistical base of AI and IS need to understand for better understanding of AI and IS.

Normal Distribution

Normal Distribution or Gauss Distribution is one of the vital elements in statistical theory and one of the vital bases for (statistical) data analysis:

- Most of natural events fit Normal Distribution.
- Most of random distributions converge to Normal Distribution.
- Most of discrete or contionous variables converge to Normal Distribution under some assumptions.

One of the most frequently use of Normal Distribution is Linear Regression. Furthermore, Principal Component Analysis and Discriminant Analysis are as an extension of Linear Regression (Koyuncugil, 2013).

Parametric Methods vs. Non-Parametric Methods

One of the main important issues in statistical data analysis is find a suitable distribution (distribution fitting) for variables. Find a suitable distribution means that variables are suitable for Parametric Methods. Otherwise, non-parametric methods or order statistics (ranks based) methods should use for data analysis. Because parametric methods require (probability) distributions and their accuracy always better than non-parametric ones (Koyuncugil, 2013) parametric methods mostly prefer instead of non-parametric ones.

Usually Good of Fit Tests (Kolmogorov-Smirnov, Chi-square etc.) use to identify the suitable probability distribution. Then the suitable parametric methods use due to the probability distribution. In case, the data fails to distribution fitting, then suitable non-parametric methods should use for analysis or modeling.

Non-Linearity

Most of the multivariate statistical data analysis which are using as a base for machine learning, deep learning, artificial intelligence and big data analytics methods based on linear methods such as Linear Regression, Principal Component Analysis and Linear Discriminant Analysis. In case lack of linearity conditions, the nonlinear modeling should use for data analysis. Currentl, one of the most popular ways to analyze non-linear data is Neural Networks. Mainly, Neural Networks Method uses activation functions such as Sigmoid Fuction which is given in Figure 3 (Berson, Smith, & Thearling, 1999; Koyuncugil, 2006). Furthermore, Sigmoid Function covers Linear Regression in case of modeling linear data. On the other hand, in case understand the data is linear before apply the NN then it is possible to develop a Linear Regression Model with better solutions than the NN Model. As a result, the most important fact is to apply the suitable method to the suitable data.

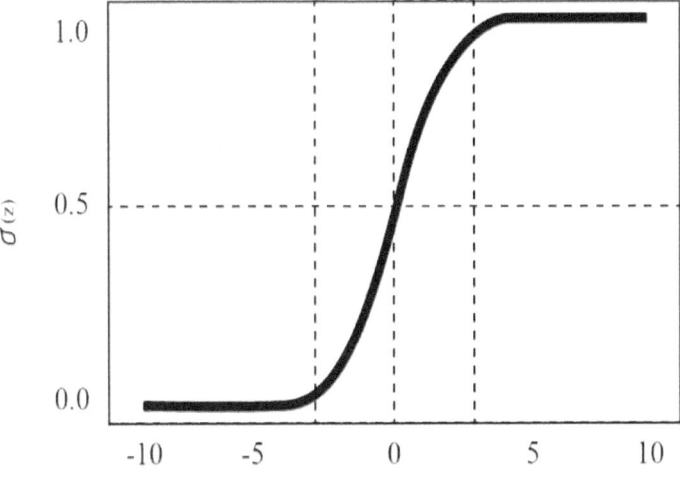

Figure 3: Sigmoid Function

One of the other ways for modeling non-linear data is piecewise linearity approach. Mainly, piecewise linearity aims to model the non-linear data with linear pieces. Piecewise linear example for Normal Distribution is given in Figure 4.

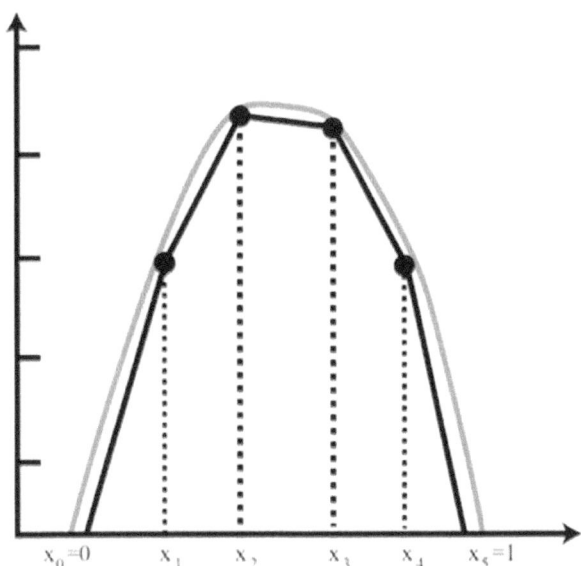

Figure 4: Normal Distribution and Its Piecewise Linearity Approach

Supervised Methods vs. Unsupervised Methods

One of the most confused concepts for data analysis is using supervised or unsupervised methods. It is possible to make different explanations for this concept but its simply about whether we can determine the output zone before analysis or not. In case the output zone can determine before the analysis then we can call it supervised. Otherwise, it is unsupervised (Koyuncugil, 2006).

We can easily use Cluster Analysis for determine the distinction between supervised and unsupervised methods. In K-means Cluster Analysis we know the data will have K clusters before we begin the analysis. So, the K-means Analysis can easily identify as Supervised. On the other hand, we couldn't know the number of clusters before we begin Hierarchical Cluster Analysis. So, the Hierarchical Cluster Analysis can identify as Unsupervised.

Mainly, unsupervised methods can use to identify/explore the data and then the supervised methods can use to verify and model the data.

Predictors

Statistical Learning Theory or in other words **Learning from Data** aims to predict non-observed data based on past data. In every data analysis we have to find statistically significant predictors for predicting the better non-observed data (Hastie, Tibshirani, & Friedman, 2001; Koyuncugil, 2006).

STATISTICAL ROOTS OF INTELLIGENT SYSTEMS

Machine Learning

We may define Machine Learning as programming the machines to solve a problem with a sample data. Programming means developing a predictive model for predicting the non-observed data based on past (sample) data (Koyuncugil, 2013; Koyuncugil, 2006). A classification diagram of Machine Learning Techniques is given in Figure 5 (The MathWorks, 2019a).

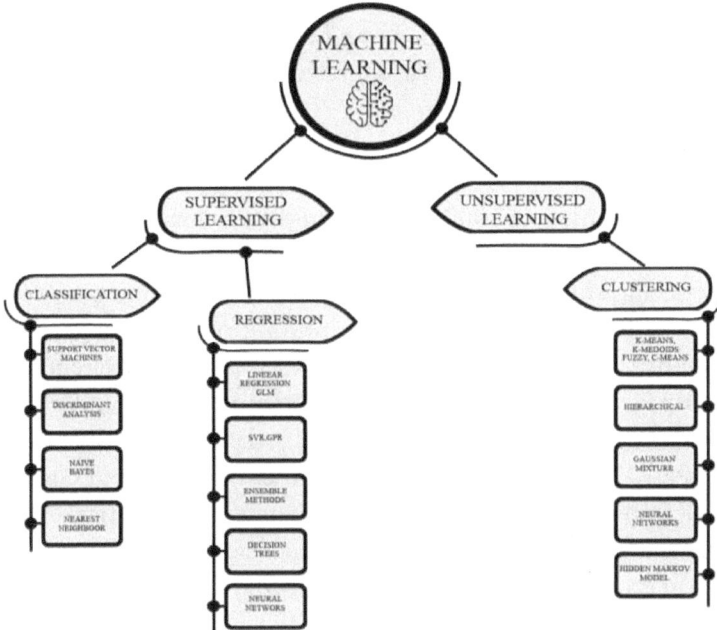

Figure 5: Machine Learning Classification

When we consider the Figure 5, we may easily see that we are talking about the some of the Multivariate Statistical Data Analysis methods. Most of the machine learnig methods are the modifications of the sub-sets of Multivariate Statistical Methods.

Deep Learning

Nowadays, one of the most popular methods is Deep Learning after availability of high performance computational devices such as GPU (or TPU) cards. Mainly, image recognition accuracy makes Deep Learning popular. Mainly, Deep Learning is a sub-set of Machine Learning which is using for big data with a high performance computers. One of the main distinctions between the Machine Learning and Deep Learning is to identification of the interactions between the variables. It is easy to identify the relation between variables in machine learning. On the other hand, many unknown computational units (hidden layers) are using to model the relation between input and output for better prediction (The MathWorks, 2019b). However, when we are talking about the Deep Learning, it means that simply we are talking about

Neuaral Networks which is based Non-Linear Statistical Modeling or in general sequencial use Machine Learning Techniques which are Multivariate Statistical Data Analysis Techniques.

Big Data Analytics

At the beginning of 2000's increasing data assumed as a chance and sequentially databases, data marts and data warehouses were developed for processing and valuating the data for better (strategic) decision making. On the other hand, at the beginning of 2010's huge, flowing data from different sources (such as social media) from different formats (such as image, video, voice) became a curse and a new understanding developed beyond the (structured/ SQL) databases as File Systems (mainly calls NoSQL). As a result, all intelligent analytics modified as big data analytics with some minor changes in Statistical Analysis, ML, DL and AI techniques mainly in data manipulations (replication, duplication etc.) for prevention of loss of data.

Data Mining

From Knowledge Discovery Concept since 1976 until now all evolutioned statistical analysis, ML, DL, AI methods gathered under one umbrella as data mining for analyzing big data for strategic decision making (Fayyad, Piatetsky-Shapiro, & Symth, 1996; Koyuncugil, 2006; Koyuncugil & Ozgulbas, 2010; Koyuncugil, 2019).

ARTIFICIAL INTELLIGENCE

Mainly, Artificial Intelligence (AI) and computer sciences have the same root planted by Alan Mathison Turing (Turing, 1936). Furthermore, Turing developed a methodology for determining whether a computer shows an intelligence which calls **Turing Machine** (Turing, 1936). Thefore, AI idea began at the same time with computer sciences development.

Artificial Intelligence is simply defined that an artificial presence (computer) which shows intelligence. It means that a computer/ machine which makes a decision by itself on an unobserved case (Koyuncugil, 2019).

Most popular AI methods are given below (Chen, Jakeman, & Norton, 2008):

- Case Based Reasoning,
- Rule Based Reasoning,
- Artificial Neural Networks,
- Genetic Algorithms, and
- Fuzzy Systems.

When we review the AI methods it is clearly emphasize that AI is based on predictions and is strongly related with (multivariate) statistical methods/theory and probability methods/theory.

Case Based Reasoning

Using similarity measures which are using in Statistical and Probability Theory mainly, case based reasoning methods are using past data (experiences) for prediction.

Case Based Reasoning Methods

- **Regression models**

$X = (X_1, X_2, ..., X_p)$ is an input vector. Y is an output value that is want predicted. So, the formula of the regression model is

$$f(X) = \beta_0 + \sum_{j=1}^{p} X_j \beta_j$$

In the formula, β_j's are unknown parameters or coefficients and X_j variables (Hastie, Tibshirani, & Friedman, 2001; Koyuncugil, 2006).

- **K – nearest neighbour**

This method is memory based and doesn't need a model to fit. x_0 query point is given and closests ones from x_r, r=1,...,k training point is find and classified to more items included in x_r. Figure

6 shows that the query point x_0 is belong to O due to 7 – nearest neighbourhood (Hastie, Tibshirani, & Friedman, 2001; Koyuncugil, 2006).

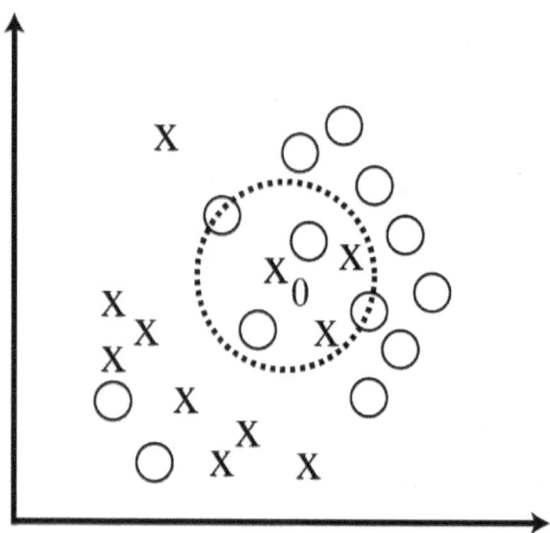

Figure 6: Nearest Neighbour of x_0 Point

Usually Euclid Distance is using because of its simplicity:

$$d_i = \|x_{(i)} - x_0\|$$

Principal Component Analysis

Principal Component Analysis is a special form of Linear Regression analysis.

Linear Regression Analysis formula as $f(X) = \beta_0 + \sum_{j=1}^{p} X_j \beta_j$, reformed via define an f(Y) function. So, new form of the formula explains X variables in Y components. PCA is used for dimension and data reduction (Ye, 2003; Koyuncugil, 2006).

K-means Cluster Analysis

K-means cluster analysis one of the most popular cluster analysis which is use an iterative assignment algorithm. It is using when all the variables are quantitative which are measurable in squared Euclid dissimilarity measure:

$$d(x_i, x_{i'}) = \sum_{j=1}^{p} (x_{ij} - x_{ij'})^2 = \left\| x_i - x_{i'} \right\|^2$$

K- means cluster analysis aims to create homogenity inside clusters and heterogenity between clusters (Hastie, Tibshirani, & Friedman, 2001; Koyuncugil, 2006).

Hierarchical Cluster Analysis

Hierarchical Cluster Analysis works like Dendogram shows in Figure 7. On the bottom each observation is individual clusters and on the top all observations in one cluster. Hierarchical cluster analysis covers the similar data from top to bottom with the distance measures such as Euclid metric (Hastie, Tibshirani, & Friedman, 2001; Koyuncugil, 2006).

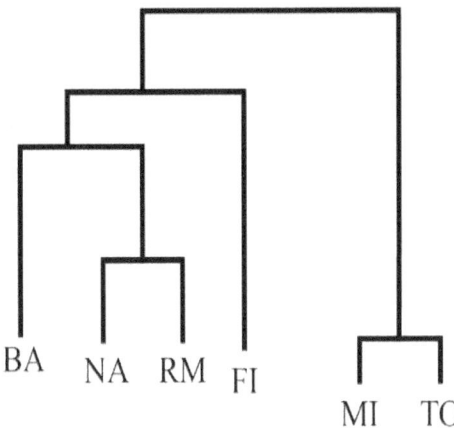

Figure 7: Dendogram

Decision Trees

On of the most up to date and popular AI methods like Neural Networks are Decision Trees. It is branching data from top to bottom via Linear Regression in C&RT Algorithm or via Chi Square Independency Analysis in CHAID Algorithm. Figure 8 shows a Decision Tree model (Berson, Smith, & Thearling, 1999; Koyuncugil, 2006).

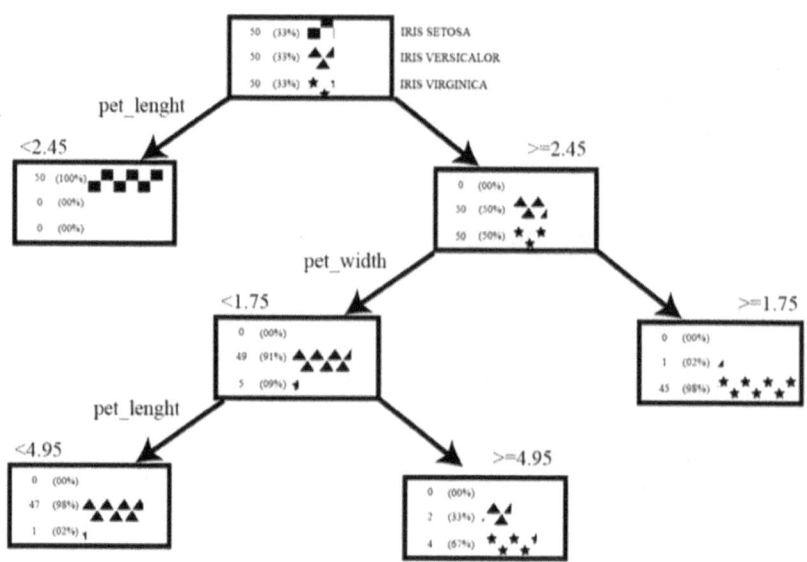

Figure 8: Decision Tree Model

Rule Based Reasoning

Rule Based Reasoning is using if-then rules in different forms. Mainly it uses Conditional Probability Theory. Rule sets inducted due to associations.

Some major rule based reasoning methods;
Association Rules:

- Market Basket Analysis and
- A-priori Algorithm.

Artificial Neural Networks

It is a non–linear modeling method which was discussed in **Non-linearity** heading under **Some Vital Statistical Elements**.

Figure 9 shows one hidden layer (Z), feed forward neural network. X's show input variable, Y's show output values and Z's show computation units (Koyuncugil, 2006).

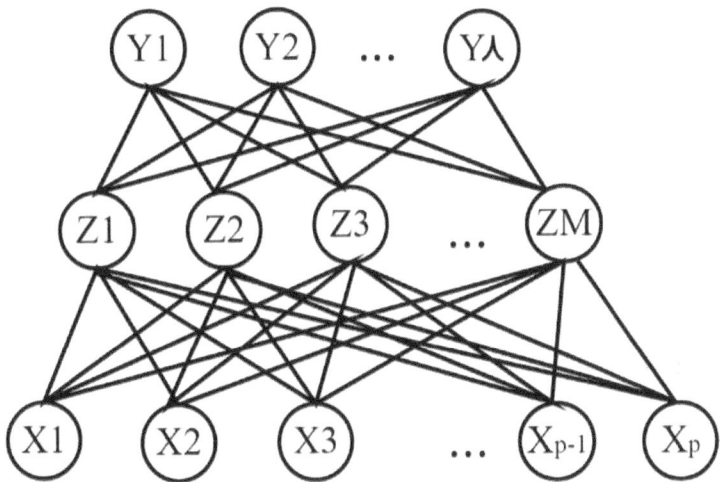

Figure 9: One Hidden Layer, Feed Forward Neural Network

Genetic Algorithms

A search technique based on sampling which is based chromosomes behaviours.

Fuzzy Systems

(Ordinary) Probability Theory assigns values as $\{0,1\}$ while Fuzzy Set Theory assigns values $[0,1]$ via Membership Functions.

Membership functions turn fuzzy values in to ordinary values. A certain (ordinary) set membership is $\forall x \in X : \mu_A(x) \in \{0,1\}$, while a

fuzzy set membership is

$$\forall x \in X : \mu_A(x) \in [0,1] \text{ (Zadeh, 1965; Koyuncugil, 2006)}.$$

Intelligent Systems

Intelligent systems are the systems including AI or the systems are showing reasoning presence. Figure 10 shows process of AI which is a base of an Intelligent System. Mainly, figure shows that the automation of a decision process as selecting necessary data, using suitable methods and making decisions via reasoning.

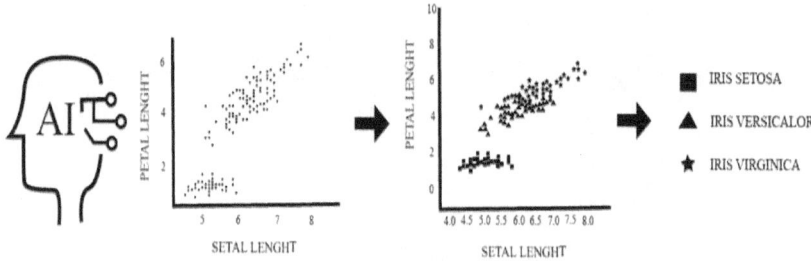

Figure 10: Artificial Intelligent Process

Examples for Intelligent Systems or AI Based Systems

Pattern Recognition Systems

Pattern is a very widely concept such as

- Voice,
- Image,
- Face.

which are barely define and uncertain (Koyuncugil, 2013).

Human mind recognizes that patterns and define them. Pattern recognition is a process of recognition by machines (computers) via pattern recognion methods.

Fingerprint recognition, hand writing recognition, satellige image recognition are the most common pattern recognition systems in our daily lives. They usually use cluster analysis and k-nearest neighbor methods for recognition.

Object Recognition Systems

It is mainly based on pattern recognition methods and image processing. Object recognition systems are the systems recognize the objects via comparing and matching objects in databases.

Facial Recognition Systems

It has almost the same working logic with object recognition systems but more sophisticated. Facial recognition systems are the systems recognize the faces via comparing and matching faces in databases. Facial recognition systems trying to match pre-defined points at the first facial recognitions systems until 2016. After 2016, deep learning methods are using with the help of GPU cards. So, computational approaches have changed and recognition speed and accuracy increased. For example, face recognition systems comparing up to 300 points on human face for recognition until 2016 but deep learning based face recognition systems are making billions of computations for face recognition after 2016.

Autonomous Systems

Autonomus systems are self driven systems such as cars, robots, drones etc. which are trying to reach pre-defined targets by their self decisions. Autonomuus systems are trained and test known cases and it continues to train itself for unknown cases via predictions based on past data (experiences). Most important factors for learning systems in general and especially autonomous systems are training rates. Less trained systems can't make accurate decisions and over trained systems can't learn new things. Best learning rates can define via train and test.

Swarm Systems

Swarm systems are self driven systems which are including more than one item such as cars, robots, drones etc. which are trying to reach pre-defined targets by their self decisions and communicating with each other independently and individually.

Conclusion

In this study, bases, definitions, methods and examples of Artifical Intelligence and Intelligent Systems are presented in details. It reveals that adequate multivariate statistical data analysis knowledge is a must for an accurate intelligent analytical model development which is the base of Artificial Intelligence and Intelligent Systems.

References

Berson, A., Smith, S., & Thearling, K. (1999). *Buildind data mining applications for CRM*. USA: McGraw Hill.

Chen, S.H., Jakeman, A.J., & Norton, J.P. (2008). Artificial Intelligence techniques: An introduction to their use for modeling environmental systems. *Mathematics and Computers in Simulation, 78,* 379–400.

Fayyad, U., Piatetsky-Shapiro G., & Symth, P. (1996). From data mining to knowledge discovery in databases. *AI Magazine, 17*(3), 37-54.

Hastie, T., Tibshirani, R., & Friedman, J. (2001). *The elements of statistical learning; data mining, inference and prediction*. Springer Series in Statistics. USA: Springer.

The MathWorks, Inc. (2019a). *What is machine-learning?* Retrieved from https://www.mathworks.com/discovery/machine-learning.html on 01 23, 2020.

The MathWorks, Inc. (2019b). *What is deep learning?* Retrieved from https://www.mathworks.com/discovery/deep-learning.html on 01 23, 2020.

Koyuncugil, A. S. (2006). *Fuzzy data mining and its application to capital markets* (Unpublished Doctoral Dissertation). Ankara University, Ankara, Turkey.

Koyuncugil, A. S. (2013). *Lecture notes for data mining course.* Retrieved from http://www.koyuncugil.org/dosyalar/vmders.pdf. on 01 23,2020. Başkent University, Ankara.

Koyuncugil, A. S. (2019). Artificial intelligence and Industry 4.0 transformation perspective in digital innovation of branding. In *International Ankara Brand Meet Up*, 20-30 November, Ankara, Turkey.

Koyuncugil, A.S., & Ozgulbas, N (Eds.) (2010). *Surveillance technologies and early warning systems: Data mining applications for risk detection.* USA: IGI Global.

Turing, A.M., (1936). On computable numbers, with an application to the entscheidungsproblem. *Proceedings of the London Mathematical Society,* 2(42), 230–265. Reprint in B. J. Copeland (Ed.) (2004), *The essential turing: Seminal writing in computing, logic, philosophy, artificial life plus the secrets of enigma.* Oxford University Press.

Turing, A. M (1950). Computing machinery and intelligence. *Mind,* 59(236), 433. Reprint in B. J. Copeland (Ed.) (2004), *The essential turing: Seminal writing in computing, logic, philosophy, artificial life plus the secrets of enigma.* Oxford University Press.

Ye, N. (2003). *The handbook of data mining.* London: Lawrence Erlbaum Associates Publishers.

Zadeh, L. (1965). Fuzzy sets. *Information and Control, 8,* 338-353.

About Author(s)

ALI SERHAN KOYUNCUGIL is Projects Coordinator of Başkent University and is a professor at Insurance and Risk Management Department. He obtained his B.Sc., Ms and PhD degrees and Associate Professor title in statistics. Prof. Koyuncugil is chairman of Artificial Intelligence Society; Elected Member (Fellow) of the International Statistical Institute (ISI); Chairman of ISI-IASC Committee on Computaional Statistics, Data Mining and Knowledge

Discovery; Member of Board of Directors of IASC (International Association of Statistical Computing); member of IASS (International Association of Survey Statisticians), Turkish Statistical Association, Turkish Informatics Society and was former vice chairman of Turkish Statisticians Association. In addition, he is Member of Advisors Board for Young Businessmen Association of Turkey. He has been taking part in a lot of international and national projects (UN, IBRD, EU, etc.). He has been taking part in a lot of international and national journals, conferences as an editorial board member, organizer, reviewer and advisor. His latest researches are on artificial intelligence, big data, deep learning, statistical learning, machine learning, pattern recognition, data mining, early warning systems, risk detection systems, association rules, facial recognition systems, object recognition systems and voice recognition systems. He has best paper award on Early Warning Systems in 2008, World's Number One in Publication on Artificial Intelligence, Data Mining, and Association Rules in 2013 according to Elsevier Scival Spotlight Index and has several best paper awards. Furthermore, Prof Koyuncugil is chairman of International Artificial Intelligence Conference and Fair to be held on October 2020 in İstanbul.

NERMIN YENIKOSE is a professor of finance. Prof. Dr. Nermin Yenikose is Vice Rector at Baskent University in Turkey. She received her licence degree in Business Management, MS degree in Health Institutions Management and PhD degree in Business Management. Her research area and publication activities include finance, cost accounting, risk management, early warning systems, information and decision-support systems used by data mining, artificial intelligence on SMEs, capital markets, health care organizations, health systems and social security. She has been taking part in many International and National projects as Coordinator, Advisor and Expert. She is also managed the "Intelligent Financial Early Warning System Project" with the established company by her as a woman entrepreneur which was granted by Government as a Research and Development Project. Prof. Dr. Nermin Yenikose established the Başkent University Gender Research Center in 2019. She conducts women's projects under this center. Prof. Dr. Yenikose has been participating a lot of international and national conferences as an organizer, reviewer and advisor.

9

TEXT MINING AS A RESEARCH AND DECISION MAKING TOOL IN SYSTEMS

SUAT ATAN

> *"I must have a prodigious amount of mind; it takes me as much as a week, sometimes, to make it up!"*
> *– Mark Twain*

ABSTRACT

Text mining is a powerful toolbox for reading huge amounts of texts like news, reports, social media. Even if it has a lot of success stories like collaborating with the prediction of future or summarization of massive tweets, it still cannot replace the reading by humans. The skill of interpretation and understanding implicit messages is peculiar to humans. Nevertheless, some features cannot be replaced by humans. Humans cannot read a million pages of texts or cannot index it about the existence of some special keywords. In this section, these aspects of text mining have been defined. Known opportunities have also evaluated.

INTRODUCTION

Checkland (2000) argues that systems thinking can be a "meta level language and theory" in which problems in many disciplines can be expressed and solved, thereby "helping to develop the unity of science". Although this view of Checkland has not been implemented yet, new tools and approaches are constantly being developed in the systems approach. One of the key burdens encountered in the transfer of systems thinking to the application area is related to the rapid change in this phase.

According to Zadeh (1962), the speed of change in activity and the changing availability of resources radically affect the reactions of the system. Therefore, there may be a discrepancy between the practical life modeling of systems thinking and the effectiveness of dynamics in the field of application. In order to overcome this incompatibility, systems thinking tools should be fed more dynamically with field information. This can be achieved by both feed forward and feedback mechanisms. Crowder, Carbone, and Fries (2019) suggest that artificial intelligence can be used as a tool to overcome this problem. The ability of systems to absorb the dynamism in the environment will improve the harmony of practical systems thinking between design and implementation. Text mining fosters the ability to design and implement based on real data in operating the forward and feedback mechanisms of systems thinking.

This section describes an approach to how text-mining tools can be used in forward and feedback mechanisms, which are indispensable basic tools for systems thinking. In the following sections, an approach is introduced to the use techniques of this tool and how it can be used as part of systems thinking.

A SHORT STORY AS A STARTING POINT

The FinanceX Corporation has used to obsess into customer data, market trends and also never-stopping finance news. This focus provided updated and incredible insight about the markets. However, this approach sometimes made workers and decision makers confused. Because the market data is tabular and there are validation methods for it. Everybody could remember the last stock price and an overall average of the market because monetary amounts are calculable and hot but the news is not. Not only the news but also the reviews of columnist, important political declarations about the economy, academic papers and expert reports was the part of "focus portfolio" it can easily be missed.

In popular science magazines and some reports, they have heard some tools which 'mines' the text. The decision makers treated these ideas like 'newfangled' think up. Like other buzzwords, big data, artificial intelligence or block-chain the 'text mining' have looked next thing which will be store big but impossible mental gadgets for tech speeches. However, one day a decision maker missed the important expert reports about the tax regulation when he returned back from his prolonged vacation, he discussed this with his assistant. The assistant complained there are at most 250 full-paged entries to read in every single day.

Decision maker decided to just try a text mining tool. The big investment in buying new software is not necessary at first. At first, he wants to enroll a data science online course, but he taught it would take time and this kind of decision just a trick of his 'deep brain' to postpone the idea. He called his friend Avinash and asked clearly: "I have 250 full-text reports to read. I remember a graph you have shown me, what is the name?" Avinash said: "Word cloud". He remembered that a word cloud is a cloud image generated from the most frequent word in all the text corpora:

Figure 1: A Word–cloud Generated from the Text of This Chapter by the author

221

If a word is bigger and has stronger font it means this word is being discussed more than other words relationally. Avinash said: "You have to code to generate it but there is an online tool you can develop it without code -of course- if you have not millions of documents. You said you have just 250 documents". Decision maker merged the files and copied all text then pasted to the online word cloud tool. The word cloud is generated. So what?

At first, he dissatisfied. There are three names of the company: Excon, Metcon and Seacon... and tons of other words. These concepts are already known by an expert in the market. The shift has finished. Decision maker cleaned his desk and shut down his computer then has gone. The next day, he tried the same thing to the new report. He started to wait for the loading graphic. Excon, Meton, MareNostrum Corp. Why Mare Nostrum Corp? He has goggled: "Mare Nostrum today". The news was striking: Mare Nostrum capital increase.

Decision maker noticed that the increasing trends of the terms about the Mare Nostrum company stems from **a logic**. The market and **hidden hand** have started to focus on the company because there is important financial news about it. Only thing was not the seeing most frequent company names, sometimes some words like **bankrupt** or **fraud** or **crisis** are also starting to appear in the word cloud.

Decision maker understands there are a lot of professional tools to analyze the market. He decides to use text-summarization tool to see the whole picture. Reading by the humans continues to be a critical part of financial analyst toolbox however text mining has become the kind of navigator for the analysts.

What is Text Mining?

Before defining the theoretical definition of text mining, it is important to clear the literal meaning of it: The term of mining requires a "valuable and huge amount of any source". The value of the text is indisputable especially if there is a collection of texts refer as **corpus**. If our source is **text** adjective of **hugeness** requires the clear-cut definition of a threshold. If the corpus is easy to read and analyze **mining** is not required. If reading every single document of the corpus is difficult or impossible the corpora is huge. Let's say one million pages of the document are huge for one person because there is no time and energy to read it no matter how valuable is. At this point, mining is necessary. So, we can say: Text mining is the collection of methodology, approaches and algorithms to analyze a huge amount of texts with the purpose of summarization, meaning extraction, sentiment extraction and other automated tasks normally doing by humans.

Opportunities of Text Mining

Employing the machine to understand the text even if not in the human level provides a lot of opportunities. The web includes tons of material written by texts and according to the analyses, 80% of the data stock of overall human-being is in the textual form. The news, blog entries, comments in the bottom of the news or blogs, reports, social media entries like tweets are commonly known sources of text mining.

Finance and Accounting: Predicting future performance of the company (Hu & Xue, 2018), to support the calculation of risk and security (Cockcroft & Russell, 2018), prediction of stock market movements (Elagamy, Stanier, & Sharp, 2018), determining economy policy uncertainty by special index, development of early warning system for business finance and economy (Klopotan, Zoroja, & Meško, 2018), quality assessment of internal audit G (Boskou, Kirkos, & Spathis, 2018), discovering bank risk factors from their financial statements (Wei, Li, Zhu, & Li, 2019), gender prediction of financial customers (Pejić Bach, Krstić, Seljan, & Turulja, 2019). These applications based on extracting text from various sources like news, analyst reports, financial statements and evaluating the extracted information with other quantitative measures like stock movements, political uncertainty or risk factors. Other studies like gender predictions just based on the texts and purpose of these studies extracting latent information from the text and descriptive statistics. The focus of finance is quantitative metrics however discussion about the finance is based on texts resources. The main problem is the volume and wide distribution of these texts. The collecting the textual data from various sources is just the first step of text analysis. The second step is dealing with this huge amount of textual data within the challenges and limitations of text mining.

Marketing: Traditional and social media have endless discussions about products, services, and brands. The people tend to reach his complain or gratitude about a company via these open platforms. All interaction between sides is in the free form. The practitioners and academicians interest these resources with motivation like the finance industry. A text mining researcher on marketing has more plentiful resources than finance because finance is a narrower field than marketing. There are millions of comments about a specific product or general mention of special industry. Analyzing market structure of sedan cars and diabetes drugs (Netzer, Feldman, Goldenberg, & Fresko, 2012), existence in the social media (He, Zha, & Li, 2013), for development of marketing intelligence tool, opinion mining about brands (Mostafa, 2013), extracting consumers needs for new products (Yoon, 2012), automated tool for research customer reviews, predicting e-commerce company success by mining the text of its publicly-accessible website (Kara, Acar Boyacioglu, & Baykan, 2011).

Strategic management: In the strategic management field text mining some studies use text mining as a research tool for bibliometric examinations: One research provides an outline of the international strategic management literature from 2000 through 2013 via 736 articles mined with text mining tools (White, Guldiken, Hemphill, He, & Sharifi Khoobdeh, 2016). Identifying new business areas is one of the strategical decisions of growing companies. Using patent information and employs a hybrid approach based on text mining and data envelopment analysis (Seol, Lee, & Kim, 2011), a similar approach but just based on text mining is performed by (Seol vd., 2011; Teichert & Mittermayer, 2002) to identify a high-technology trend. Frequency of words is one of the most powerful tools of the text-mining toolbox and looks useful to catch the buzzwords and determine the next technology. However, Yoon (2012) uses a different point of view by focusing on weak signals. He employs text mining to catch weak signals of long-term business opportunities.

These opportunities cited from the business administration field are examples that limited by their field and base textual corpus. Opportunities for the usage of the text mining tools are not bounded with the social sciences. It can be employed wherever text matters. For instance, mining e-mails are one of the practical application for the organizations (Tang, Pei, & Luk, 2014) to visualize workload, e-mail categorization (Bogawar & Bhoyar, 2012), spam detection. There is some software solution to perform email mining. Text mining can help healthcare applications (Jensen, Jensen, & Brunak, 2012; Popowich, 2005) too. Mining textual records about health are also an example of usability of the text mining.)

Challenges and Limitations of Text Mining

The main challenge of text mining is dealing with the complexity of human languages: Grammatical exceptions, typos, uncategorizable definitions, implicit explanations, and other properties. For example, word frequency is common metrics to understand the topic of any corpora. Let's say the word **crisis** is repeated more than 1000 times in the analysis. We can interpret that the crisis is a common word and reflect the theme of the corpora. However, we should have a suspect: Is the term **crisis** repeated in every single of a document or it is just mentioned a few marginal documents. Another saying, is the word crisis homogenous or heterogeneous? This information is important because interpreting the corpora depends on. If the term homogenous we can easily say the crisis is a common theme vice versa. There is another metrics TF.IDF to solve this homogeneity problem. However, the challenges of text mining are still overwhelming.

TEXT MINING METHODOLOGY

Definitions

Document: The document means any single integrated text block constitutes one or more sentence. A document may be news, blog entry, tweet or a paragraph of the book. A document reflects a meaningful piece of information. **Corpus:** The collection of the document constitutes of documents. Let's say we are doing a text mining analysis on special financial newspaper. The collected news collection is naming as corpus. The document and corpus may seem like a table and database in traditional databases. The different tables comprise the database. **Term**: A term or word constitutes sentences and a sentence or more than one sentence constitutes the document. Although the sentences are unique it is constituted by the limited list of terms because every language has a dictionary and terms are limited. Whereas, there may be special names like place or brand names at those terms should be meaningful and interpretable. **Stop-words:** Natural language has some terms which have a secondary role within the construction of sentences. The auxiliary verbs or other elements like that have not special meaning when they are alone. These terms named as stop-words. In text mining stop-words typically are being removed.

Basic Analysis

A text-miner starts to processes the given corpus. This process includes stop-words removal, lowercasing the terms, non-alphanumeric characters (like parenthesis and punctuation) removal. After preprocessing the text-miner prepares Document-Term-Matrix (DTM) or Term-Document-Matrix (TDM). This matrix is a two-dimensional matrix which columns come from every unique term from the corpus. The rows reflect documents and typically sign by the unique ID of the documents. A value crossing in the term-document pair reflect repeat of a given term in given document ID. Most frequent words, bi-grams and other extraction from the text are from the DTM or TDM like digitized words.

To elaborate the how the collection of documents being presented by numbers and quantitative metrics here a simple illustration: Let's say there is a news collection includes just five documents. To define clear and followable example supposing our analysis just focuses on the title of news instead of their full text.

Table 1: A Corpus Example

Document ID	News Story Title
1	The best streaming apps for *kids* for **Netflix**
2	**Netflix** has a strategy for Chinese movies and Chinese Language
3	**Netflix** Signs Deal with Alibaba to Add Chinese-Language TV Show
4	**Netflix** buys StoryBots as competition for *kids'* shows heats up

In this collection (or corpora) every word in every document can be extracted and listed to show which word is in which document. Naturally, repeating words like **for** or another frequent word can be seen in a different document. Therefore, there is no need to restate these words. Showing repeating of words within the document just by digits which reflects a count of repetition is enough.

Table 2: Document Term Matrix of Example Corpus

Document Id	best	stream	app	kids	Netflix	strategy	Alibaba	movie	Chinese
1	1	1	1	1	1	0	0	0	0
2	0	0	0	0	1	1	0	1	1
3	0	0	0	0	1	0	1	0	1
4	0	0	0	0	1	0	0	0	1

The matrix at the above is named as a document-term matrix (DTM). The first row this matrix is the same as the news list, columns are extracted unique words from the overall collection. Some words may detract from the word list due to their weak or zero meaning. The numbers under the word columns reflect existence and repetition. Zero means the relevant word is not found in the intersecting document ID. One reflects existence. The numbers more than one show how many times the relevant word is repeated in the document with relevant ID. For instance, the term **Chinese** repeated two times in the second document.

Without additional calculation, DTM matrix may show a lot of details about the collection. Before showing the details let's show how it works. To summarize calculation over mathematical notation on this matrix would be useful. A document is D_i in the collection $C = \{D_i \mid (i = 1, 2, ...n)\}$. A number in the DTM matrix reflects the repeat of a word (j) in the document ID i is $d_{i,j}$. For example, $d_{3,9} = 2$ and this reflects the word 'Chinese' repeated in document 2. The DTM matrix a matrix has dimension ixj.

Most frequent words list

The sum of columns in the DTM or $\sum d_j$ for each column is a vector consisted of numbers. This vector reflects the sum of repetition of each word reflected in the column. The $\max\left(\sum d_j\right)$ is the most frequent word of the whole collection. If we sort this vector from biggest to smallest number, this list will give the list of most frequent words in the whole collection. For our example when we calculate row sums, we get $\sum d_j = \{1,1,1,2,3,1,1,1,1\}$ this set is mapped the words: j [best, stream, app, kids, Netflix, strategy, Alibaba, movie, Chinese]. When we sort the list according to the number of frequencies, the most frequent words should be the word **Netflix** with 3 frequencies. The second is **kids** with 2 frequencies.

In this basic example, we just show the process of computation for DTM. For bigger collections and longer documents, DTM becomes very large. This kind of huge matrices cannot be managed by a desktop spreadsheet application. Therefore, the programming languages like R and Python are being used to manage this kind of big matrices and another type of datum. A collection may consist of millions of documents. Millions of documents are not readable by anyone or by any team. In such cases, even extracting the most frequent words is useful to understand what the collection keeps in itself.

N-grams and co-occurrence analysis

Phrases which include two or more words are a typical component of any language. However above-mentioned analysis is based on breaking sentences into word by word. What would we do when we want to extract phrases like **TV Shows**. This is a two-word component but reflects a single meaning. When we want to break this kind of components down we should do *n-gram analysis*. N-gram is a linguistic component which comprises of n words. When n=2 it also refers as **bigram** or n=3 is **trigram**. The computation of most frequent n-grams is a bit complicated than the extraction of most frequent words. However, there are embedded functions in the R and Python language text mining libraries to compute this feature. N-grams functioning like most frequent word lists but they may provide more information than most frequent word lists.

Before explaining the co-occurrence analysis, we should emphasize that an n-gram is comprised of two words are absolutely together. To elaborate it more here is an example: In the sentence "Chinese-Language TV Show" **TV Show** is n-gram because these two words are together. Otherwise let's address this sentence "In a TV channel, a show named". In this sentence the word **TV** and **Show** in the same document they are not together. Therefore, in this case, the words TV and show do not constitute an n-gram. Occuring of two words in same document is called co-occurrence. If two different words (not in the bigrams) occurred frequently across the collection we can sort the most co-occurred words. Let's think about economic news. All news includes the word **finance** also most probably include the word **crisis**. In that case, there is a red flag to set a relation between finance and crisis and focus on to hypothesis like: "There is a sign of financial crisis". The co-occurrence analysis provides this feature.

Sentiment Analysis

Sentiment analysis purpose is to classify a huge amount of separate document. The classification depends on the aim of the research. Typically, sentiment analysis performing for classification of texts or tweets is labeled according to the positive, negative and neutral. For instance, sentiment analysis may focus on the classification of the tweets about a politician. Deeper sentiment analyses are also possible. The labels may be extended more detailed interpretation like angry, enjoy hope, enthusiasm, etc.

Although sentiment analysis task leans on text mining classification, the subtask of prediction of any single document may be done with various approaches. Clearest and easiest way works the positive and negative word lists (dictionaries). The algorithm gets every word of the document and matches the positive and negative word lists then matching words is being summed up. The positive and negative scores being calculated with these matching words and algorithms decide the positivity and negativity.

Typically, a sentiment analysis can be performed in two ways. First is dictionary based or lexicon based analysis. This analysis based on selective checking of predefined sentiment dictionaries. These methods roughly work as on checking every word in the document and comparing it whether that word positive or negative and counting positive and negative terms. The dictionary can be expanded. For instance, any researcher can get or prepare a word list which includes words about anger or stress or happiness also possible. The second method of sentiment analysis does not require any dictionary. It based on machine learning (ML) techniques. In this sort of analysis, an ML algorithm requires a data set like Netflix news in our example but with more columns. The additional column should include the human assessment of each document (news). A sophisticated ML algorithm process this information gain a predictive capacity for new documents (news).

Application of dictionary-based analysis easier than ML algorithm and does not require any ex-ante human assessment to train the ML algorithm. Accuracy of this algorithm based on the selection of the sentiment dictionaries. Polysemic words are cuss of sentiment dictionaries. Also, a typo in the dictionary based analysis stems from unmatched words and weak analysis. Flexibility is less because of ever-changing linguistic terms and new buzzwords.

The machine learning approach is more flexible because it is trained by humans rather than word selection. Accuracy of the ML algorithm can be improved by known ML techniques. There are tons of ML algorithm and their extensions, therefore there is not only one ML algorithm for sentiment analysis. From traditional techniques, Naive Bayes, Decision Tree and new ML techniques like Deep Learning and Convolutional Neural Networks are common algorithms for sentiment analysis on texts. These algorithms are also powerful middles to classify the texts according to their content.

Machine Learning

Machine learning is a new field which stands in the intersection of mathematics, computer science, and statistics. The most concrete task of machine learning is a prediction. Any machine algorithm requires the data to do this task. The prediction of any case with a given data set may be a classification or guessing the one or more property of unknown observation. Let's suppose there is a dataset which includes news titles and categories of this news. These categories are labeled by humans to help the web page visitors find the relevant news according to their interest. In this case, each new news item should be label by a human. However, a machine learning (ML) algorithm can do this task instead of the humans. The algorithm discovers the pattern of news via their including words then generates a **predictor** model to be able to predict new unlabeled news. There is a lot of different ML algorithm to perform this task. Every algorithm has a different statistical and computational background. To empower accuracy and also the speed of these algorithm studies are continuing. Any ML researcher can measure the accuracy of this algorithm by separation part of labeled data to benchmark with labeled by the predictor model.

To explaining how machine learning works let's use the same example of Netflix News. Machine learning requires labeled data. The **labeling** each row of news or any type of data. Initially, the **labeling** activity is being performed by humans to use this data training of machine. For our example let's consider the labeled data as below:

Table 3: Labeled Corpus Example

Document ID	News Story Title	Label
1	The best streaming apps for *kids* for **Netflix**	Positive
2	**Netflix** has a strategy for Chinese movies and Chinese Language	Positive
3	**Netflix** Signs Deal with Alibaba to Add Chinese-Language TV Show	Neutral
4	**Netflix** buys StoryBots as competition for *kids'* shows heats up	Negative

When there are enough rows and their labels, a machine learning algorithm can derive patterns from the text. When any predictive model is generated this model can label any new observation (for our example new news story) according to the generated model. The generated model is being generated the given training data. Once the algorithm predicts, the calculation of the accuracy of this model is also possible. The accuracy calculation is to understand the power of predictive model and benchmark with other models. So, how machines extract this kind of patterns without any external manipulation and just with data. While there are a plethora of machine algorithm and their explanations, we just focus to explain the algorithm that its algorithm is less sophisticated. The Naive Bayes algorithm is based on Bayesian Possibility. Roughly, the Naive Bayes algorithm calculates the labels according to the multiple combinations of the possibilities. For example, let's contemplate an obsessive deaf employer who can just read but cannot hear the names of fruits in the gross. It prepares a repetitive table like below:

Table 4: An Example for Features of Fruits

Shape	Color	Name of Fruit
Long	Yellow	Banana
Long	Yellow	Banana
Sphere	Orange	Grapefruit
Sphere	Orange	Orange
Sphere	Orange	Orange
Long	Yellow	Banana

It is easily discernible that only fruit which long and yellow is always banana. However, if it is sphere and orange there are two possible outcomes: orange or grapefruit. The Naive Bayes algorithm is sure because there is strict consistency of banana without extension. How would the algorithm differentiate orange and banana? It that case the population is matter. If most of orange and sphere things our training data is orange the algorithm says a new undefined thing is a likely orange. Of course, the real data set is not so simple. Let's consider every word of our Netflix example as properties like shape, color and etc. and the repeat of the words as the values. In that case, for instance, any news includes the **crisis, bankrupt** and does not include **success, solution** may always be labeled as negative. The Naive Bayes would learn on the principle of fruit example.

The sentiment analysis and machine learning algorithm both provide an automated classification of texts. These tasks may be more than two classes (label). For instance, a collection of a news story may be labeled like the economy, sport, politics, magazine, health, etc. In positive-negative classification and news category classification, eacsh news item gets just one label. However, by multinomial algorithms like KNeighboors, one observation (news) can be labeled more than one. Another saying finding relevant tags to news is also possible. Besides, automated classification is provided language detection either.

When machine trained by English, French, and Bambara, Hindu and Tamil language it gets the power of distinct these languages when any e-mail being received the system will label the email as its language and can send to relevant persons who can understand this message.

Semantic Networks

When two words are together either over by bi-grams or co-occurrence these two words and their relation can be defined as a network. The words comprise **nodes**, and relationship comprises **edge**. The network visualization and other metrics of network analysis may provide further insights about the text corpus. Therefore, the networks which extracted from the words are defining as a semantic network. To explain it more, let's show an example. We use some of the co-occurrence words from our Netflix example and visualize it with Python NetworkX library. We color red when a word (node) has more than one repetition. Python codes and its result are self-explanatory:

```
import network as nx
G = nx.Graph()
G.add_node("kids",repeat=2)
G.add_node("netflix",repeat=4)
G.add_node("strategy",repeat=1)
G.add_node("app",repeat=1,)
G.add_node("chinese",repeat=3)
G.add_node("movie",repeat=1)
G.add_node("language",repeat=1)
G.add_node("alibaba",repeat=1)
G.add_edges_from([
                ('kids','netflix'),
                ('kids','app'),
                ('netflix','app'),
                ('netflix','movie'),
                ('netflix','strategy'),
                ('strategy','movie'),
                ('chinese','language'),
                ('netflix','language'),
                ('strategy','chinese'),
                ('netflix','alibaba')
                ])
```

```
#colorizing according to the frequency
#https://stackoverflow.com/questions/27030473/how-to-set-colors-for-nodes-in
-networkx-python
color_map = []
for node in G:
    #https://stackoverflow.com/questions/13698352/storing-and-accessing-node-
attributes-python-networkx
    node_repeat = G.node[node]['repeat']
    if node_repeat > 2:
      color_map.append('red')
    else: color_map.append('orange')

import matplotlib.pyplot as plt
pos=nx.circular_layout(G)
nx.draw(G,node_color = color_map,with_labels = True, pos = pos)
plt.show()
```

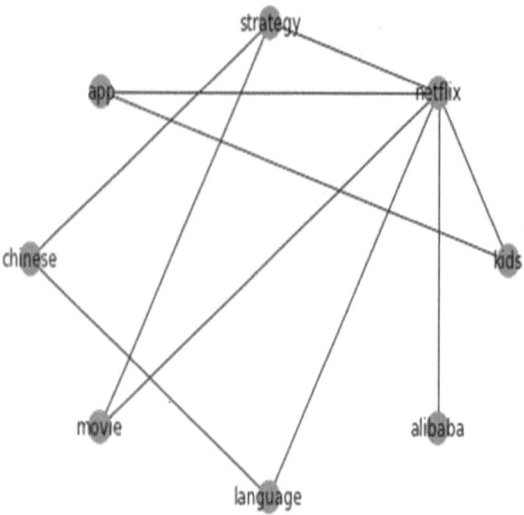

Figure 2: Result of Code Block 1

When we look at our semantic network carefully, we easily see the word **Netflix** have more connection than others. Therefore, the code set its color red due to its frequency. This visualization explains the semantic relationship the words. Any words which often passed different documents get more connection due to its co-occurrence. For these kinds of small networks, the visualization may not explain something new however; if corpus gets to grow the graph will tell more.

Other Methods

Text Mining includes more method than we mentioned here (for instance, summarization, part-of-speech recognition, etc). Summarization is the task of automated summarizing the long texts and corpus. By this tool, any researcher can gather general information about the topic and substantial theme of the huge amounts of textual assets. Part-of-speech recognition is the task of automated detecting the linguistic elements of sentences. The Natural Language Processing (NLP) sub-field deals with this task. By the opportunities that NLP provides the nouns, adjectives and adverbs can easily be extracted. The verbs are no exception. Not only the verbs but also their tenses can be extracted from the text. The text generation algorithms should be mentioned here: Any text generator algorithm like LSTM can learn to set new sentences by the learning any corpus. For instance, a trained LSTM algorithm from the corpus of tweets of a special person(s) can tweet new ones via mimicking their wording and rhetoric. Another method is semantic networks. Semantic networks are a visualization of the n-grams or phrase of terms in 'n' item long. By the visualization of n-grams, the most common text phrases and their directions may be visualized.

Text Mining Tools

Text mining tools can be categorized as a Graphical User Interface (GUI) based tools and command line (CL) based tools. Usage of GUI based tools like IBM SPSS Modeller, SAS Text Miner and Weka does not require coding the steps. Every step during the analysis is performed by the classical menu and form items. On the other hand, the CL based tools like R and Python programming languages requires familiarity with coding and algorithm development. All these tools perform main text mining tasks however some of them may have special tools to enrich the analysis. For instance, the R language has a special package (igraph) to visualize any network. This feature provides visualization of the semantic networks. Another way of categorization is the licensing of this software Link. Generally, CL based tools like R and Python are open source and absolutely free however the GUI based tools are commercial applications except for Weka. Weka is an open source project developing by University Waikato from New Zealand.

R and Python have useful libraries/packages for text mining tasks. A library/package is a kind of module of these languages which developed for specific tasks. When any researcher needs to special task like text mining, he or she can access the open-source documents and tutorials of relevant library/package which are well and centrally documented in the web [9]. These libraries/packages are not directly embedded in the relevant language for productivity and speed. The store of these libraries is accessible within applications like RStudio. When any library/package is being required it easily installs from libraries/packages store. In that time all functions are ready to use. In the same project, it is possible to use more than one library/package.

This modularity and flexibility give incredible usefulness for both languages and their tools. Any package can contemplate like special software. The modularity gives developing pipelines between special packages. For instance, with to package in the R the researcher can mine any kind of text. Besides, the **igraph** package is dedicated to network analysis. The researcher can import both packages and can use the output of functions of the **tm** package as the input of the **igraph** package for semantic network visualization. Developing machine learning models also possible with this modularity there is a lot of machine learning libraries within the R packages.

Any text mining task requires a huge amount of texts. If it is available above-mentioned tools can help. However, if there is no textual data in the hand, obtaining these texts may be a problem. The web includes a plethora of textual sources like news outlets, user reviews, blogs, etc. Getting these sources by hand is not possible. For this problem R and Python, languages provide special packages for fetching and extracting text from the web. The **harvest** package in the R provides tools for fetching texts from any kind of websites. The fetching the text from web page to build a text corpus is naming as 'scraping'. Python also has an advanced library **scrapy** for big scraping projects. When any web page scraped, cleaning the irrelevant parts

---

9   For instance, text mining (tm package) of R: https://cran.r-project.org/web/packages/tm/tm.pdf

(advertisements, images, texts or the menu item) and extracting text nugget from it requires special effort. For these tasks, R and Python also provide special extractors. There is some API's like **diffbot** for more professional text extraction.

REFERENCES

Bogawar, P. S., & Bhoyar, K. K. (2012). Email mining: A review. *IJCSI International Journal of Computer Science Issues, 9*(1), 429-434.

Boskou, G., Kirkos, E., & Spathis, C. (2018). Assessing internal audit with text mining. *Journal of Information & Knowledge Management, 17*(02), 1850020.

Checkland P. (2000). Soft systems methodology: A thirty year retrospective. *Systems Research and Behavioral Sciences, 17*, 511-558.

Cockcroft, S., & Russell, M. (2018). Big data opportunities for accounting and finance practice and research. *Australian Accounting Review, 28*(3), 323-333.

Crowder, J. A., Carbone, J., & Friess, S. (2020). Systems-level thinking for artificial intelligent systems. *Artificial Psychology,* 15-27.

Elagamy, M. N., Stanier, C., & Sharp, B. (2018). Stock market random forest-text mining system mining critical indicators of stock market movements. In *2nd International Conference on Natural Language and Speech Processing (ICNLSP),* 1-8.

He, W., Zha, S., & Li, L. (2013). Social media competitive analysis and text mining: A case study in the pizza industry. *Int J. Information Management, 33*, 464-472.

Hu, N., & Xue, F. (2018). Forward-looking statement and corporate future Performance—Based on Text Mining and Machine Learning Technology *Available at SSRN 3289215.*

Jensen, P. B., Jensen, L. J., & Brunak, S. (2012). Mining electronic health records: Towards better research applications and clinical care. *Nature Reviews Genetics, 13*(6), 395-405.

Kara, Y., Acar Boyacioglu, M., & Baykan, Ö. K. (2011). Predicting direction of stock price index movement using artificial neural networks and support vector machines: The sample of the Istanbul Stock Exchange. *Expert Systems with Applications, 38*(5), 5311-5319.

Klopotan, I., Zoroja, J., & Meško, M. (2018). Early warning system in business, finance, and economics: Bibliometric and topic analysis. *International Journal of Engineering Business Management, 10*, 1847979018797013.

Mostafa, M. M. (2013). More than words: Social networks' text mining for consumer brand sentiments. *Expert Systems with Applications, 40*(10), 4241-4251.

Netzer, O., Feldman, R., Goldenberg, J., & Fresko, M. (2012). Mine your own business: Market-structure surveillance through text mining. *Marketing Science, 31*(3), 521-543.

Pejić Bach, M., Krstić, Ž., Seljan, S., & Turulja, L. (2019). Text mining for big data analysis in financial sector: A literature review. *Sustainability, 11*(5), 1277.

Popowich, F. (2005). Using text mining and natural language processing for health care claims processing. *SIGKDD Explor. Newsl., 7*(1), 59-66.

Seol, H., Lee, S., & Kim, C. (2011). Identifying new business areas using patent information: A DEA and text mining approach. *Expert Systems with Applications, 38*(4), 2933-2941.

Tang, G., Pei, J., & Luk, W.-S. (2014). Email mining: Tasks, common techniques, and tools. *Knowledge and Information Systems, 41*(1), 1–31.

Teichert, T., & Mittermayer, M.-. (2002). Text mining for technology monitoring. In *IEEE International Engineering Management Conference, 2,* 596-601.

Wei, L., Li, G., Zhu, X., & Li, J. (2019). Discovering bank risk factors from financial statements based on a new semi-supervised text mining algorithm. *Accounting & Finance, 59*(3), 1519-1552.

White, G. O., Guldiken, O., Hemphill, T. A., He, W., & Sharifi Khoobdeh, M. (2016). Trends in international strategic management research from 2000 to 2013: Text mining and bibliometric analyses. *Management International Review, 56*(1), 35-65.

Yoon, J. (2012). Detecting weak signals for long-term business opportunities using text mining of Web news. *Expert Systems with Applications, 39*(16), 12543-12550.

Zadeh, L. (1962). From circuit theory to system theory. *Proceedings of the IRE, 50*(5), 856-865.

About Author(s)

Suat Atan is a software developer in the Agricultural and Rural Development Institution (ARDSI). He received his degree PhD from Ankara University with the thesis on text mining on the financial news of Turkey Stock Exchange Market. He has worked various in companies and public institutions including the Turkish Treasury. Since 2009, he has been working at ARDSI. His interest includes studies on text mining and machine learning. He has two published books on Google Cloud Platform and Data Analysis on R Language.

INDEX

www.ingramcontent.com/pod-product-compliance
Lightning Source LLC
Chambersburg PA
CBHW020739180526
45163CB00001B/282